NATURAL ENVIRONMENT RESEARCH COUNCIL

INTERNATIONAL HYDROLOGICAL DECADE

Hydrological Research in the United Kingdom (1970-1975)

Natural Environment Research Council
Alhambra House, 27-33 Charing Cross Road, LONDON, WC2H 0AX

Steep stream flume, Severn catchment, Plynlimon

PREFACE

The Natural Environment Research Council includes hydrology among the subjects for which its Charter gives it powers to encourage, support and carry out appropriate research. Council also has a responsibility for research training which is provided by Research Grants and Training Awards to the universities for students reading for higher degrees. The first edition of this booklet was published by NERC half way through the International Hydrological Decade (1965-1974). Its aim was to outline the hydrological interests and expertise of all Government-sponsored agencies in the United Kingdom. In the intervening period since its publication there have been many changes in both the water industry and in the executive departments which, together with an increase in undergraduate and postgraduate support in the universities, reflect the growing awareness of the importance of hydrology to national affairs.

It is hoped that this second edition provides the information relevant to current hydrological progress in the UK in a concise and helpful fashion. Many people have co-operated in the compilation of this report and their assistance is gratefully acknowledged.

INSTITUTE OF HYDROLOGY
Crowmarsh Gifford
Wallingford
Oxon OX10 8BB

November 1975

CONTENTS

TABLES

INTERNATIONAL HYDROLOGICAL DECADE

An earlier edition of this publication – Hydrological Research in the United Kingdom (1965-70) – was published by the Natural Environment Research Council (NERC) as a UK contribution to the International Hydrological Decade (IHD) to coincide with the mid-Decade conference. The British National Committee for the IHD was pleased to be associated with that earlier addition, which was welcomed internationally as well as in the UK by hydrologists and others concerned with water resources.

The British National Committee has been enthusiastic in encouraging NERC to publish this edition covering the second five years of the IHD; it is hoped that the publication will prove as useful as the earlier edition.

The first five years of the IHD were notable in the UK as a period during which hydrological research and technology received special impetus from the Water Resources Board (WRB) created by the Water Resources Act of 1963 and from NERC set up under the Science & Technology Act of 1965. The initial support given to hydrological investigations and research by these new bodies was evident in the earlier edition from the information given by various UK agencies and by individuals including university workers. This encouragement has continued during the second five years of the Decade, and effort and expenditure on UK hydrological research has increased steadily. As before, this research has been sponsored or carried out by various Government and other agencies in addition to the WRB and NERC, and in the universities. An important feature of the later part of the period covered by the present publication has been a reorganisation of central and regional management of water supplies and sewage disposal in England and Wales as described in Chapter I.

Internationally, it has been agreed that the IHD should be succeeded by an International Hydrological Programme (IHP) supported by the United Nations Educational Scientific and Cultural Organization (UNESCO) and by the World Meteorological Organization (WMO). The British National Committee for the IHD has contributed to the formulation of this proposed programme and has supported its adoption in the Co-ordinating Council of the IHD. In looking ahead to effective implementation of the IHP it is hoped that the present publication will be useful in reviewing research in progress and its relevance to future research that may be included in the Programme.

A E SEDDON
Chairman, British National Committee
International Hydrological Decade

British Technical Aid activity, Iran *Institute of Geological Sciences*

CHAPTER I

DEVELOPMENT AND ORGANISATION OF HYDROLOGICAL RESEARCH IN THE UNITED KINGDOM

As with every other commodity, consumer demand for water has increased rapidly in recent years. Stimulated by the upsurge in user requirement, Parliament passed the Water Resources Act of 1963 which created the River Authorities with responsibility for water conservation and the Water Resources Board to advise both them and central Government on such matters.

But providing and maintaining wholesome and adequate water supplies, coupled with looking after and cleaning up our rivers, is a complex and interrelated affair. It needs large-scale investment to budget the more than adequate national water income in such a way that it is in the right places at the right time and of the right quality. Recognition of the enormity of the task was reflected by the creation of the Department of the Environment in 1970 from the Ministries of Housing & Local Government, Transport, and Public Building & Works. This Department took over the responsibility for established research organisations such as the Hydraulics Research Station, the Water Pollution Research Laboratory and the Water Research Association and also set up a Directorate General Water Engineering to advise both Government and the industry on water supply, waste water treatment and disposal. The Directorate also sponsors research and development projects on these topics.

Next, the weakness in the organisation of the water industry itself, particularly the separation between quantity and quality, was corrected with the passing of the Water Act 1973. This provided for the creation of ten regional water authorities to take over the existing functions of river authorities, water undertakers, sewerage and sewage disposal authorities; these new bodies now have responsibilities for water resources in the widest sense.

The influence of environmental factors affecting the natural water balance must be taken into account in regional planning, a need which had been recognized some years ago and resulted in the setting up of the Hydrological Research Unit in 1962 by the Department of Scientific and Industrial Research. The Unit was given the task of carrying out research into the effect of land use on the water balance within catchments. Previously, hydrology had been relatively neglected as a subject for research although limited funds for university research had been provided by the same Department.

The Unit (which became the Institute of Hydrology in 1968) was subsequently transferred to the Natural Environment Research Council set up under the terms of the Science and Technology Act 1965. The Council's Royal Charter gave it the authority to assume responsibility for the development of the science of hydrology and it was empowered to provide funds for hydrological research not only in the United Kingdom but overseas. Expenditure on hydrological research from Council funds falls broadly under three headings:

(1) The research programme of the Institute of Hydrology.

(2) The research programme of the Hydrogeological Department of the Institute of Geological Sciences.

(3) Research grants and training awards for universities.

The expenditure in the year 1974/75 under these three heads was £1,347,000.

The recent reorganisation within the water industry has been paralleled by changes in the organisation and funding of Government-sponsored hydrological research. This was initiated firstly by the Government's decision published in a White Paper* in 1972 to transfer substantial funds from the Science Vote to executive departments, following Lord Rothschild's proposals for customer/contractor research management; secondly, by the creation of the Water Research Centre from the Water Research Association, the Water Pollution Research Laboratory and parts of the Water Resources Board and the Directorate General Water Engineering. Such complete reorganisation sets in motion mechanisms for closer contact between research workers and water engineers, important for the satisfactory utilization of the country's water resources.

The extent and range of current research topics is shown in the tables of Chapter II.

*Framework for Government Research and Development, *Cmnd* 5046

Clywedog reservoir dam

SYSTEMATIC TABULATION
OF
HYDROLOGICAL RESEARCH

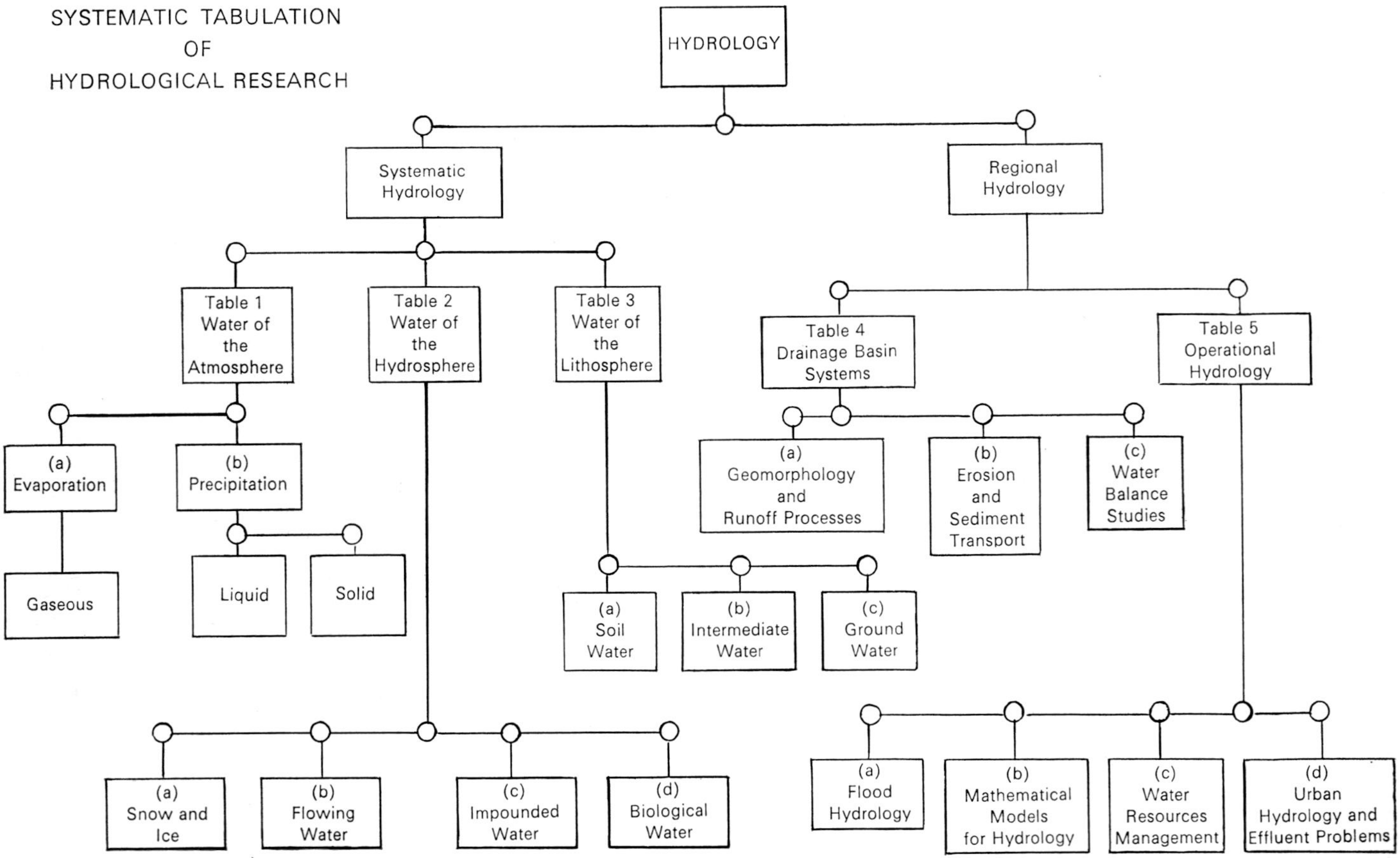

CHAPTER II

TABULAR STATEMENT OF RESEARCH

The separations and sub-divisions of scientific hydrology used as a basis for classification of research topics are shown in the diagram opposite. There are two major groupings, namely: Systematic Hydrology and Regional Hydrology. The research projects in the first group have been assembled under general sub-headings within the tables where appropriate and, as in the previous edition, allocated to one of three columns describing the nature of the study, i.e. *Instruments and Techniques*, *Physical Hydrology and Experimental Work* and *Theoretical Studies and Data Analysis*.

Every research project is labelled with the organisation or organisations responsible, using the abbreviations listed overleaf. Further information about any particular project can be obtained from Chapters III and IV where the research interests of government agencies and university departments are described in more detail.

KEY TO ABBREVIATIONS USED IN TABLES 1 – 5

Abbreviation	Organisation
ABDBT	Aberdeen University – Botany Department
ABDGG	Aberdeen University – Geography Department
ABDGL	Aberdeen University – Geology and Mineralogy Department
ABDSS	Aberdeen University – Soil Science Department
ADAS	Agricultural Development and Advisory Service
AL	Appleton Laboratory
AERE	Atomic Energy Research Establishment, Harwell
ASTCE	Aston in Birmingham University – Civil Engineering Department
AWA	Anglian Water Authority
BAS	British Antarctic Survey
BKLGG	London, Birkbeck College – Geography Department
BMNRPB	Banff, Moray and Nairn River Purification Board
BQUGG	Belfast, Queen's University – Geography Department
BRE	Building Research Establishment
BRMCE	Birmingham University – Civil Engineering Department
BRMGG	Birmingham University – Geography Department
BRMGL	Birmingham University – Geology Department
BRMPH	Birmingham University – Physics Department
BRDCE	Bradford University – Postgraduate School of Studies in Civil Engineering
BRPCE	Brighton Polytechnic – Civil Engineering and Building Department
BTLCE	Bristol University – Civil Engineering Department
BTLGL	Bristol University – Geology Department
CAMGG	Cambridge University – Geography Department
CAMPH	Cambridge University – Physics Department
CRPB	Clyde River Purification Board
CTYCE	City University – Civil Engineering Department
CWPU	Central Water Planning Unit
DENI	Department of the Environment for Northern Ireland
DGWE	Directorate General Water Engineering
DUNCE	Dundee University – Civil Engineering Department
DUNGL	Dundee University – Geology Department
DUNME	Dundee University – Mechanical Engineering Department
EANCR	East Anglia University – Climatic Research Unit
EANEV	East Anglia University – School of Environmental Sciences
EDGFN	Edinburgh University – Forestry and Natural Resources Department
EDGMT	Edinburgh University – Meteorology Department
EMRS	East Malling Research Station
EXEGG	Exeter University – Geography Department
FBA	Freshwater Biological Association
FC	Forestry Commission
GLACE	Glasgow University – Civil Engineering Department
GRI	Grassland Research Institute
HRS	Hydraulics Research Station
HULPB	Hull University – Plant Biology Department
HULGG	Hull University – Geography Department
IGS	Institute of Geological Sciences
IH	Institute of Hydrology
LANEV	Lancaster University – Environmental Sciences Department
LL	Letcombe Laboratory
LARS	Long Ashton Research Station
LDSGG	Leeds University – Geography Department
LGC	Laboratory of the Government Chemist
LIVGG	Liverpool University – Geography Department
ICLBT	London, Imperial College – Botany Department
ICLCE	London, Imperial College – Civil Engineering Department
ICLGL	London, Imperial College – Geology Department
KCLCE	London, King's College – Civil Engineering Department
KCLGG	London, King's College – Geography Department
MAFD	Ministry of Agriculture Field Drainage Experimental Unit
MANCE	Manchester University – Civil Engineering Department
MANGG	Manchester University – Geography Department
MISR	Macaulay Institute for Soil Research
MNICE	Manchester University Institute of Science and Technology – Civil and Structural Engineering Department

KEY TO ABBREVIATIONS USED IN TABLES 1 – 5 (*continued*)

Abbreviation	Organisation
MNIPH	Manchester University Institute of Science and Technology – Physics Department
MO	Meteorological Office
MIPGG	Middlesex Polytechnic – Geography and Planning Department
NCAFE	National College of Agricultural Engineering – Field Engineering Department
NEWCE	Newcastle upon Tyne University – Civil Engineering Department
NEWGG	Newcastle upon Tyne University – Geography Department
NIDA	Northern Ireland Department of Agriculture
NOTCE	Nottingham University – Civil Engineering Department
NOTSA	Nottingham University – School of Agriculture
NSHEB	North of Scotland Hydro-Electric Board
NVRS	National Vegetable Research Station
NWA	Northumbrian Water Authority
NWWA	North West Water Authority
OXFAS	Oxford University – Agricultural Science Department
OXFFY	Oxford University – Forestry Department
OXFGG	Oxford University – Geography Department
PLPES	Plymouth Polytechnic – Environmental Sciences Department
QMCGG	London, Queen Mary College – Geography Department
RDGAB	Reading University – Agricultural Botany Department
RDGGG	Reading University – Geography Department
RES	Rothamsted Experimental Station
SALCE	Salford University – Civil Engineering Department
SHECE	Sheffield University – Civil and Structural Engineering Department
SHEGG	Sheffield University – Geography Department
SHEGL	Sheffield University – Geology Department
SIAE	Scottish Institute of Agricultural Engineering
SNPCE	Sunderland Polytechnic – Civil Engineering Department
SNPGL	Sunderland Polytechnic – Geology Department
SOTCE	Southampton University – Civil Engineering Department
SOTGG	Southampton University – Geography Department
SSEW	Soil Survey of England and Wales

Abbreviation	Organisation
STAGG	St Andrews University – Geography Department
STRAP	Strathclyde University – Applied Physics Department
STRCE	Strathclyde University – Civil Engineering Department
STRGG	Strathclyde University – Geography Department
STWA	Severn-Trent Water Authority
SURCE	Surrey University Engineering Department
SWA	Southern Water Authority
SWWA	South West Water Authority
TRRL	Transport and Road Research Laboratory
TWA	Thames Water Authority
TWPCE	Teeside Polytechnic – Civil Engineering Department
UCLGG	London, University College – Geography Department
UCLGL	London, University College – Geology Department
USP	Unit of Soil Physics
WABGG	Wales: University College, Aberystwyth – Geography Department
WAS	Regional Water Authorities
WDU	Water Data Unit
WISCE	Wales: Institute of Science and Technology, Cardiff – Civil Engineering and Building Technology Department
WNWDA	Welsh National Water Development Authority
WPBS	Welsh Plant Breeding Station
WPRL	Water Pollution Research Laboratory
WRC	Water Research Centre
WSWCE	Wales: University College of Swansea – Civil Engineering Department
WSWGG	Wales: University College of Swansea – Geography Department
WWA	Wessex Water Authority
YWA	Yorkshire Water Authority

TABLE 1 WATER OF THE ATMOSPHERE — (a) EVAPORATION

Instruments and Techniques		Physical Hydrology and Experimental Work		Theoretical Studies and Data Analysis	
		Measurement of total evaporation			
IH Development of automatic weather station and evaluation of prototype commercial equipment	**SWWA** Evaluation of solar radiation integrator as part of automatic weather station	**MO** Comparison of various evaporimeters (Kew Observatory and other sites)	**NOTSA** Evaporation from a catchment		
IH Use of Fluxatron to measure energy budget directly	**WWA** Assessment of automatic climatological station at Chew Stoke	**LDSGG** Radiation budget of a small pasture catchment			
		MAFD Studies of spring time soil drying processes as affected by artificial drainage			
				Estimation of evaporation	
				MO Computer model for estimating evaporative losses	**BRDCE** Evaporation studies
				MO Comparison of different methods of assessing evaporation	**MO** Quality control of evaporimeter data
				MANCE Evaporation estimation in a stratified atmosphere	**MO** Studies of Penman estimates of potential evaporation
				CWPU MO Evaluation of long-term catchment evaporation	

TABLE 1 WATER OF THE ATMOSPHERE — (a) EVAPORATION (*continued*)

Instruments and Techniques	Physical Hydrology and Experimental Work	Theoretical Studies and Data Analysis
	Measurement of evaporation from vegetation	
MO Experimental evaporimeter and simple irrigated lysimeter (Kew Observatory)	**IGS** Comparative studies of infiltration and evaporation losses using large undisturbed drainage lysimeters isolated laterally by membranes **IH** Evaporation characteristics of tall vegetation **AWA** Direct measurement of evapotranspiration **GRI** Drainage lysimeters; comparison with Penman estimates and evaporation tank	
		Estimation of spatial and temporal variability of evaporation
		LARS Effects of altitude, aspect and slope on evaporation rates **STWA** Evaluation of spatial variations in evaporation over Severn basin **MO** Evaporation in London area since 1697 **STRAP OXFGG** Spatial and temporal variability of potential evaporation
	Effects of plants on transpiration	
	RES IH Plant and soil factors in evaporation **ABDBT** Carbon dioxide and water vapour exchange in a forest canopy	

TABLE 1 WATER OF THE ATMOSPHERE — (a) EVAPORATION *(continued)*

Instruments and Techniques	Physical Hydrology and Experimental Work	Theoretical Studies and Data Analysis
	ICLBT IH Plant physiological influences on evaporation from forests **EDGFN** Response of trees to wind speed and humidity	
	Measurement of evaporation of intercepted precipitation **IH** Study of interception losses	
		Analysis of Data **EDGMT** Physiological and synoptic-scale influences in diurnal humidity variations over vegetation

TABLE 1 WATER OF THE ATMOSPHERE — (b) PRECIPITATION

Instruments and Techniques		Physical Hydrology and Experimental Work		Theoretical Studies and Data Analysis	
IH AL MO Applications of remote sensing techniques to hydrology	**MO** Use of radar to measure areal rainfall	**FBA MISR LGC** Chemical composition of rain water **MNIPH** Microphysics of rainfall	**AL** Small-scale structure of rainfall, using rapid-response raingauges in conjunction with radio propagation experiments		
		Spatial and temporal variability of precipitation			
	MO Trials of different raingauges, variously mounted	**AL** Drop size distribution and local variation in rainfall	**EXEGG** Spatial variations in rainfall input to a small instrumented catchment	**BTLGG** Estimation and prediction of rainfall from weather satellite data	**ICLCE MO** Rainfall storm models
MO Design of reference, gravimetric raingauge (Kew Observatory)	**LL IH MO** Magnetic tape rainfall recorders	**LDSGG** Topographic controls on the distribution of rainfall in the Central Pennines	**BRMCE** Rainfall characteristics at a point and over areas	**EANCR** Modelling rainfall in south-east England	**IH** Standard gauge catch deficiencies correction
NSHEB IH Remote recording of rainfall	**SOTCE** Performance of raingauges	**MO** Studies of rainbands and associated motions using Dopler radar (Malvern)	**QMCGG** Meso-scale investigation of frontal precipitation	**MO NSHEB** Relationship between point and areal rainfall	**MO** Extreme and probability analysis of rainfall on different time and space scales
NSHEB Evaluation of tipping bucket raingauges and punched tape recorders	**NSHEB IH NWWA** Exposure of raingauge at ground level	Effects of orography on rainfall distribution		**MISR** Three maps assessing climatic conditions in Scotland based on potential water deficit and other parameters	Forecasting quantitative rainfall on different time and space scales
NWWA Development of Bestel Dean interrogation and alarm raingauges		Studies on synoptic and meso-scale precipitation mechanisms	**SOTGG** Relationship of precipitation distribution to streamflow	**CWPU MO** Evaluation of long-term catchment rainfall, monthly and daily	Mapping of rainfall by computer Improved quality control system for rainfall observations
		WSWGG Effect of changes in the frequency of synoptic types on precipitation	**NEWGG** Effective rainfall reaching a stream channel in an upland catchment	**SWA/WRB DENI/MO IH** Computer processing of rainfall data	Annual UK rainfall averages 1941-70 **ICLCE** Areal rainfall from mathematically-defined surfaces
		LANEV Small-scale structure of rainstorms			

TABLE 1 WATER OF THE ATMOSPHERE — (b) PRECIPITATION (*continued*)

Instruments and Techniques	Physical Hydrology and Experimental Work	Theoretical Studies and Data Analysis
	Crop response to rainfall	
	STRAP/IH Fog interception in coniferous forest	**NVRS** Effects of rainfall on crop response
	IH MISR Interception of precipitation by coniferous forest	**ICLBT** Modelling rainfall interception in forest canopies
		Analysis of rainfall events
		BRPCE Analysis of rainstorms over Sussex
		SHEGG Storm rainfall in northern England, 16.7.73
		CWPU MO Relationship between historical weather information and river flows to increase length of river flow records
		CWPU MO Relative severity of historic dry periods

TABLE 2 WATER OF THE HYDROSPHERE — (a) SNOW AND ICE

Instruments and Techniques	Physical Hydrology and Experimental Work	Theoretical Studies and Data Analysis
	BAS BRMPH CAMPH Physics of ice flow	**OXFAS** Changes in snow distribution in Scottish Highlands
	NEWCE Snowmelt river flooding in the British Isles	**MO** Studies of snowfall, water equivalent of snow and snow melt
		LGC Determination of concentration and biodegradability of de-icing chemicals in runoff from airfields

TABLE 2 WATER OF THE HYDROSPHERE — (b) FLOWING WATER

River flow

Instruments and Techniques

WWA
Tilting weirs as permanent gauging structures

AWA
Performance of crest tapping pressures on ERD standard weirs

AERE
Radioactive tracers for field calibration

IH WRC
Dilution gauging

SWWA
River gauging using air bubbles

WRC
Electromagnetic gauging stations

NSHEB DENI/WDU
Evaluation of punched tape recorders

SWWA
Low-cost telemetry devices

STWA
Slope discharge method for flow gauging

HRS
Performance of two-dimensional and flat-Vee triangular profile weirs

Performance of thin plate weirs

Behaviour and calibration of current meters

Field calibration of weirs

Computer programs for calibration of critical depth flumes

SHECE
Weir calibration by pondage drawdown techniques

STWA
Ultrasonic flow measurement

IH
Magnetic tape loggers

WISCE
Less conventional flow gauging

Physical Hydrology and Experimental Work

SWA
Effects of large increase in summer flow of chalk stream as part of river regulation studies

IH SHEGG EXEGG
Low flow characteristics

CWPU
Assessment of artificial changes in river flow

Losses from stream flow during river regulation

STRCE
Laboratory and numerical studies of runoff from a conceptual catchment

BRMCE
Determination of travel times in rivers

Theoretical Studies and Data Analysis

ICLCE
Stochastic modelling of long-term persistence in streamflow sequences

WISCE
Finite element analysis of flow systems

WNWDA
Extension of river flow records for planning development

ICLCE
Stochastic models of monthly streamflow

Models for synthetic data generation

ICLCE
Shot noise processes as stochastic models of daily streamflow data

SWWA
Sampling intervals for stage/flow measurement

CWPU BRMCE
Multi-site flow data generation

SWWA BRMCE
Stochastic daily flow generation

CWPU
Generation and validation of synthetic flow sequences

Flow prediction

Development of computer programs to obtain catchment regression models from streamflow archives

WSWCE
Numerical simulation of runoff

TABLE 2 WATER OF THE HYDROSPHERE — (b) FLOWING WATER (*continued*)

Instruments and Techniques		Physical Hydrology and Experimental Work		Theoretical Studies and Data Analysis	
		Hydraulics of flow			
WWA Stream bed lining	**SWWA** Performance of low cost bed controls	**MAFD** Agricultural drains; design of filters and entry slots **ICLCE** Velocity fluctuations in natural stream channels **STRCE** Hydraulic geometry of tidal watercourse	**CAMGG** Hydraulic geometry of rivers **SOTCE** Flow resistance, sediment transport and bed formation in alluvial channels **SNPCE** Ripple generation on granular bed of open channel flow	**HRS** Frictional resistance of alluvial channels **CAMGG** Theoretical models of river channel pattern Hydraulic geometry of channels of variable roughness **HRS** Mathematical prediction of water level envelope downstream of a breaking dam	**HRS** Unsteady flow in drainage channels with lateral inflow Mathematical modelling of steady and unsteady flows **NEWCE** Friction coefficients for flows with small depth and high roughness
		Water quality			
WPRL Development of portable, submersible water quality monitors Recording equipment for dispersion studies using radioactive tracers **FBA** Measure of solar irradiance above and within chalk streams **WNWDA** Evaluation of open channel fish counting on a large river	**FBA IH** Chemical analysis of river waters **HRS** Monitoring silt concentration in rivers and estuaries **STWA** Development of water quality monitors and telemetry network	**NWWA** Algal growth variation Artificial freshet releases **BTLGG** Hydrological processes affecting downstream changes in quality **EDGFN** Hydrological characteristics of fish spawning streams **STRGG** Thermal characteristics of upland streams	**FBA** Distribution of freshwater arthropods Temperature regime of chalk streams **DUNCE** Water quality on peat catchments **DENI/WRC** Quality/quantity regime, River Lagan **WRC** Effects of sewage effluent and storm-water discharge on biota and water quality of rivers	**BRMCE** Stochastic model for prediction of dissolved oxygen in polluted river **SOTCE** Mathematical models for circulation and dispersion studies in shallow waters	

TABLE 2 WATER OF THE HYDROSPHERE — (b) FLOWING WATER (*continued*)

Instruments and Techniques	Physical Hydrology and Experimental Work	Theoretical Studies and Data Analysis
	LGC Chemical and radio-chemical analysis of water supplies; special reference to trace constituents **EXEGG** Spatial and temporal variation in solute transport, Exe catchment **IH FBA** Natural water chemistry **FBA** Chemistry-macrophyte detritus system in experimental circulating stream **NWWA** Relationship of iron and coloured organic acids in peaty waters **MISR** Nutrient balance of small peat catchment **EANEV** Concentration of dissolved phosphorus in natural waters Variations in major dissolved constituents in certain East Anglian rivers **NEWCE SHECE** Hydrological significance of hydrochemistry **NWWA** Nitrate consumption in river systems **SHEGG** Solute sources and variations in streams, mid-Wales **WABGG** Solutes and sediments in mid-Wales	

TABLE 2 WATER OF THE HYDROSPHERE — (c) IMPOUNDED WATER

Instruments and Techniques		Physical Hydrology and Experimental Work		Theoretical Studies and Data Analysis	
FBA Chemistry of lakes Physical factors of lakes **EANEV** Tracer studies in the Norfolk Broads **WRC** Tracer studies of retention times in bankside storage reservoirs	**FBA** Controlling influences on phytoplankton **FBA** Methods of dating recent sediments Pollen analysis of sediments	**FBA** Changes in fauna of upper River Tees due to building of Cow Green reservoir **DENI** Paleolimnology of Lough Erne **FBA** Soil history as from chemical analysis of sediments Pollen analysis giving vegetational history of lake catchments Palaeolimnological expertise **DENI/NIDA** Effects of afforestation on water quality in lakes and reservoirs **WRC** Limnology of stored water **TWPCE** Deterioration of waters stored in flooded coal mines	**HRS** Density currents and sedimentation in reservoirs **FBA WRC** Dynamics of summer stratification in phytoplankton production Factors controlling viable microbial populations in lakes **DENI** Development of new lake (at Craigavon) in urban surroundings **NWWA** Bioassay for algal growth limiting nutrients **DENI** Assessment of nutrient loading in lakes including means of reducing phosphorus	**STWA** Planning models for Craig Gogh reservoir and Shropshire groundwater **SWA** Optimization of conjunctive use of water resources **CWPU** Combined use of rivers, reservoirs and desalination plants **WRC** Reliability of supply, for given storage capacity and input **SHECE** Flood control in reservoirs and storage pounds	**FBA** Mathematical model of water movement in lakes **MNICE** Probability aspects of reservoir operations **LGC** Chemical and radiochemical analysis of water supplies; special reference to trace constituents **DENI/NIDA** Mathematical modelling of Lough Meagh **CWPU** Resource allocation studies

TABLE 2 WATER OF THE HYDROSPHERE — (d) BIOLOGICAL WATER

Instruments and Techniques		Physical Hydrology and Experimental Work		Theoretical Studies and Data Analysis
STRAP LL Use of fibreglass lysimeters		**EMRS** Relation of osmotic and turgor pressures to tissue contraction and stomatal closure; effects on cropping	**ABDSS** Effects of water supply on growth of grass and Sitka spruce	**STRCE** Mathematical relationships of soil-water-plant continuum
	LL Use of ^{15}N in utilization of nitrogen fertilizer by crop, and leaching through soil	**NVRS** Uptake of nutrients and herbicides Examination of root systems and effect on water abstraction	**ABDBT** Photosynthesis and transpiration in Sitka spruce in relation to genotype and environment	
		OXFFY Hydrological interpretation of plant root distributions	**LL** Influence of transient water logging on plant growth	
		NIDA Effects of drainage on tree growth and stability	**FC** Growth and adaptive changes in tree roots in soils subject to water logging	**OXFFY** Water relations of forest trees
		GRI Growth of forage crops in relation to water supply – irrigation	**WPBS** Growth response of grass varieties to water stress	

TABLE 2 WATER OF THE HYDROSPHERE — (d) BIOLOGICAL WATER *(continued)*

Instruments and Techniques	Physical Hydrology and Experimental Work		Theoretical Studies and Data Analysis
	EDGFN Influences on growth of Sitka spruce **GRI** Effects of chemical and physical treatments on the loss of water from detached leaves under controlled conditions and from swaths of cut foliage	**WPBS** Crop water use of grass varieties and of lines selected for anatomical and physiological features which influence water transfer **OXFFY** Hydrological relations of forest and moorland vegetation **FC** Tree growth on chalk soils – rooting, depth of drying	

TABLE 3 WATER OF THE LITHOSPHERE — (a) SOIL WATER

Instruments and Techniques		Physical Hydrology and Experimental Work		Theoretical Studies and Data Analysis	
		Soil water content			
IH Neutron probe techniques **USP** Laboratory measurement of hysteretic soil-water relationships on soil columns **BRE** Equipment for measuring pore water pressure and permeability	**IH** Development of automatic soil station **FC** Measurement of water potentials and contents, permeability, and oxygen flux in soil profiles and spoil; using tensiometers, resistance blocks, bore-holes, and polaro-graphic oxygen gauge	**FC** Moisture retention measurements in a range of soils Measurement of changes in soil water regime following afforestation, cultivation and drainage, and clear-felling	**DUNME** Hydraulic conductivity of humified peats **SSEW** Field observations of water levels in dip-wells, in piezometers and by neutron probe; measurement of saturated hydraulic conductivity. Laboratory measurement of bulk density, porosity and moisture retention characteristics	**MO** Studies of effective rainfall, soil moisture deficits Operational model for computing soil moisture deficits and evaporation over the UK	
		Soil water movement			
EANEV Use of fluorescent dyes for tracing soil moisture movement **MAFD** Use of lysimeters in a controlled environment		**IH** Neutron probe and tensiometer use for vertical soil water fluxes in unsaturated zone **ABDSS** Water regimes in gley soils **HULPB** Water level fluctuations in peat in relation to rainfall	**LL** Size and continuity of soil pores in relation to movement of water and gases **USP** One-dimensional laboratory investiga-tions on hysteretic moisture profile development	**NVRS** Models for distribution of water in cropped and uncropped soils	**USP** Equilibrium and development of moisture profiles; rate of infiltration into and drainage from a moisture profile Theory of soil-water movement and equilibrium in terms of physical properties of soil; hysteretic effects Green and Ampt approach in soil-water movement studies

TABLE 3 WATER OF THE LITHOSPHERE — (a) SOIL WATER (*continued*)

Instruments and Techniques	Physical Hydrology and Experimental Work		Theoretical Studies and Data Analysis
	SHEGG Effects of vegetation on runoff/infiltration ratio	**CTYCE** Thermodynamics of soil moisture and minerals	
	Infiltration characteristics of small Pennine catchment	**HULGG** Infiltration and soil moisture movement in a boulder-clay plot	
	IH Soil water process studies for physically-based models of small catchments	**AERE/CWPU IH** Recharge and movement using radioactive tracers	
	MAFD Water table levels related to soil type and artificial drainage	**RDGGG** Investigations into throughflow in soils	
		STRGG Infiltration rates and over-land flow development on various soils and land treatment	
		IH Drainage through natural piping in upland peat soils	

TABLE 3 WATER OF THE LITHOSPHERE — (a) SOIL WATER (*continued*)

Instruments and Techniques	Physical Hydrology and Experimental Work	Theoretical Studies and Data Analysis
	Soil studies and classification	
UCLGG Soil erodibility based on physical and chemical measurements	**BKLGG** Hydrometeorological and pedological factors in soil-water disposition **SSEW** Classification of soils according to drainage characteristics, winter rain acceptance poten-tial, water table fluctuations **FC** Classification of forest soils, especially uplands, by water regime, peat depth and drainage status, and correlation with tree growth, risk of windthrow, and required drain spacing	**SSEW** Recognition and definition of soil water regime types (wet and dry) by field and laboratory studies of soil profiles and related samples Linear extensibility of fine textured soils **MISR** Moisture tension, release characteristics, storage capacity and other parameters of mineral and organic soils **ADAS** Determination of soil moisture characteristics
	Regional studies	
	BRMGG Soil moisture variability in the Forest of Dean **LDSGG** Actual evapotranspiration and soil moisture, eastern Pennines	**BTLGG** Examination of patterned ground, Severn estuary

TABLE 3 WATER OF THE LITHOSPHERE — (a) SOIL WATER (*continued*)

Instruments and Techniques	Physical Hydrology and Experimental Work		Theoretical Studies and Data Analysis	
	Drainage and irrigation; crop yields			
	USP Infiltration and drainage studies	**NEWGG** Crop and soil water relationship in an agricultural catchment, Northumberland	**NVRS** Models of crop response to NPK fertilizers under different soil and weather conditions	**SOTCE** Analysis of magnitude and variation of irrigation water requirement
	ICLCE Drainage of peat lands			
	NIDA Effects of drainage on tree growth and stability, and on borehole levels	**FC** Effects of different tree species on soil water regime, especially comparisons of pines with spruces and firs		
	ABDSS Soil water regimes, with or without secondary drainage treatment	**IH EMRS** Crop water use studies with tensiometers and neutron probes. Studies of plant water status and soil moisture stress relations		
	EANEV Relationships between soil moisture retention characteristics and irrigation requirements			
	IH Water uptake by Scots pine; fluxes and storage within trees	**WPBS** Comparison of leaching and water losses on contrasting soils under native and sown grassland		

TABLE 3 WATER OF THE LITHOSPHERE — (a) SOIL WATER (*continued*)

Instruments and Techniques	Physical Hydrology and Experimental Work	Theoretical Studies and Data Analysis
	Soil water chemistry; herbicides, nutrients, etc., pollution	
	EMRS GRI Relationship of water in soil profile to availability of plant nutrients **LARS** Movement of persistent insecticides through soil. Effects of weed-free environment, achieved by herbicides, on infiltration **NVRS** Persistence of herbicides	**NVRS** Models for movement and uptake of nutrients Persistence of herbicides in the field
	Agricultural and civil engineering aspects of soil water	
	NIDA Interpretations of tree root distributions in deep peats and gleyed soils in relation to ploughing **FC** Influence of clear-felling on soil properties on clayey soils, with reference to establishing next crop **BKLGG** Agricultural betterment resulting from land drainage and for flood protection **BRE** Measurement of pore pressure and its influence on the stability and deformation of the ground, natural slopes, excavations, earthworks, foundations and dams **BKLGG** Changing surface-depression storage and infiltration opportunity in ploughlands	**BRE** Prediction of pore pressure in gravel layers underlying Thames flood embankments during storm surge conditions **ICLCE** Design of motorway drainage systems **GRI** Models of crop protection in relation to water supply

TABLE 3 WATER OF THE LITHOSPHERE — (a) SOIL WATER (*continued*)

Instruments and Techniques	Physical Hydrology and Experimental Work	Theoretical Studies and Data Analysis
	SSEW Measurement of resistance by soil cores to core penetrometers at varying suctions, water contents and bulk densities, to relate soil strength to field properties	
	MAFD LL Effects of artificial drainage on agricultural soils including drainflow, soil moisture, crop yield, etc.	
	SIAE Effects of cultivation and soil compaction on movement and retention of soil water	
	OXFAS Application of fertilizers on soil, well and streamwater	
	Field drainage and climatic change on water level	

TABLE 3 WATER OF THE LITHOSPHERE — (b) INTERMEDIATE WATER

Instruments and Techniques	Physical Hydrology and Experimental Work	Theoretical Studies and Data Analysis
IGS Hydrochemical – mainly stable isotope and trace element – techniques **IGS** Destructive and non-destructive in-situ moisture sampling techniques in unsaturated zone profiles	**Infiltration and recharge studies** **IGS** Infiltration into permeable strata by direct and indirect methods in Lower Greensand, Bunter Sandstone and Chalk. Tracer techniques in the vadose zone **IH** Monitoring recharge using natural tritium **TWA** Recharge of Lower Greensand, London basin **RDGGG** Throughflow in soils	**IGS** Analyses of rates of infiltration to permeable strata and theoretical relationships of infiltration to recharge and groundwater discharge **TWA** Estimation of recharge **CWPU** Regional assessment of infiltration to groundwater storage to aid resource studies
	Water chemistry **SWA** Infiltration mechanism of sewage effluent on Upper Chalk **SWA/WRC** Hydrogeology of nitrate transport in Upper Chalk **IGS** Behaviour of synthetic leachates in unsaturated zones of representative aquifers (Lower Greensand, Chalk and Plateau Gravels)	

TABLE 3 WATER OF THE LITHOSPHERE — (c) GROUNDWATER

Instruments and Techniques		Physical Hydrology and Experimental Work		Theoretical Studies and Data Analysis	
Aquifer properties; pumping tests; water/rock interactions					
FC Automatic monitoring of bore-hole levels using capacitance as a transducer; electrical dipstick for manual measurements	**IGS** Multicapability modular wireline borehole logging equipment	**IGS** Field investigation of three-dimensional flow in major aquifers using pumping test, borehole wire-line logging and closed circuit TV	**TWA** Determination of aquifer hydraulic properties	**SWA** Digital models of chalk aquifers	**SWWA** Aquifer modelling using digital computers
IGS BRMGL Application of geo-physical methods (seismic and electrical)	**IGS** Determination of physical properties of strata and column of water in a well (laboratory and field)	**BRMCE** Unconfined aquifer behaviour	**TWA** Assessment of borehole performance	**NEWCE** Groundwater resources	**WRC** Semi-automatic analysis of well pumping tests
SWWA Automatic package for monitoring small scale pump tests	**BRMGL** Electrical properties of Permo-Triassic sandstones		**BRMCE NWWA** Aquifer test design for pumping in vicinity of a river		**TWPCE** Non-uniform aquifer flow
IGS Physical properties of reservoir rocks – permeable porosity and pore-size distribution	**IGS/HRS** Assessment of simple inexpensive borehole screens for use overseas		**SNPGL** Aquifer potential of Fell Sandstone, Northumberland		
TWA Inexpensive well logging techniques			**OXFGG** Solution and percolation rates of selected limestones		
WRC Surface and downhole geophysical methods; borehole drilling and sampling; well pumping tests; chemical and hydraulic properties of rock cores; petrological examination of aquifer materials			**ADAS** Measurement of water-table heights in a range of soils		

TABLE 3 WATER OF THE LITHOSPHERE — (c) GROUNDWATER (*continued*)

Instruments and Techniques	Physical Hydrology and Experimental Work		Theoretical Studies and Data Analysis	
		Hydrogeology and regional groundwater studies		
	MISR Water table fluctuations in hydrological sequence of soils on granitic till	**SWWA/BRMGL** Surface geophysical investigation of Triassic aquifer of east Devon	**TWA** Electrical analogue models for groundwater studies	**TWA** Hydrogeological relations between Chalk and Upper Greensand, western London basin
	SWWA/IGS Hydrogeology and geochemistry of Triassic aquifer of east Devon	**BTLGL/SWWA** Hydrogeology of Permian aquifer of mid-Devon		
	SWA Groundwater resource studies in the Chalk of Hampshire, Sussex and Kent	**BRMGL STWA** Geophysical investigations in Shropshire Trias relating to groundwater development	**CWPU** Conservation of groundwater. Variations in availability of groundwater storage	**LANEV** Modelling groundwater problems using digital computers
	NWWA Sub-drift surface in Lancashire and Cheshire	**NWWA** Hydrogeology of Manchester area	**IGS** Analysis of water level data for predictive purposes	**CWPU** Hydrological forecasting of groundwater levels
	STWA Hydrogeology of the Bromsgrove area	**ABDGL** Hydrogeology of Knock Hill, Banffshire		**SWA** Development of a computer system for processing groundwater levels
	ABDGG Preliminary investigations of springs on Ben Rinnes	**BRMGG** Groundwater conditions in the Peak District		
	ABDGL Geophysical investigation of Alford Basin, Aberdeenshire	**BRMGL** Hydrogeological assessment of central Lincolnshire limestone		

TABLE 3 WATER OF THE LITHOSPHERE — (c) GROUNDWATER (*continued*)

Instruments and Techniques	Physical Hydrology and Experimental Work		Theoretical Studies and Data Analysis
	BRMGL Hydrogeology of Keuper Sandstone in Droitwich syncline area	**BTLGL** Groundwater studies of East Midland aquifers	
	BTLGL Hydrogeology of Jurassic limestones of Cotswolds	**ICLGL** Groundwater studies in East Anglia	
	HULGG Groundwater hydrology of a boulder clay catchment	**TWA** Chalk aquifer in Wandle and Hogsmill catchments	
	MNICE Hydrogeology of Permian and Triassic aquifers, northern England	**IGS** Comparative study of sedimentology and hydrogeology of Lower Greensand	
	SWWA/BRMGL Hydrogeology of granitic island, St. Mary's, Isles of Scilly	**IGS** Groundwater resources of Triassic sandstones of Lagan Valley, N.I.	
	UCLGL Hydrogeology of Jurassic aquifers	**HULGG PLPES** Karst water studies, Central Pennines	
	IGS Groundwater resources of Old Red sandstone, Stratheden	**UCLGL** Hydrogeology of Triassic strata – Wirral Peninsula	
	WWA Groundwater studies in Cotswolds and Mendips	**YWA** Hydrogeological investigations, Vales of York, and Pickering	

TABLE 3 WATER OF THE LITHOSPHERE — (c) GROUNDWATER (*continued*)

Instruments and Techniques		Physical Hydrology and Experimental Work		Theoretical Studies and Data Analysis	
		Input and recharge to groundwater			
		CWPU SWA WRC Artificial recharge of Lower Greensand and its development within a regional combined use system	**TWA** Regional geophysical investigations of inflow horizons	**CWPU WRC** Assessment of role of artificial recharge in regional and national groundwater development	**USP** Groundwater movement to sinks (especially agricultural drains), application of fundamental theory and use of Hele-Shaw and electric analogues
			TWA Induced infiltration of river water into Chalk	**IGS WRC** Mathematical studies of well efficiencies	**USP** Application of finite element method in numerical solutions of groundwater problems in heterogenous soils
				TWA Estimate of recharge to Thames aquifers	
		Groundwater flow			
CWPU AERE Use of radioactive isotopes to study age and movement of groundwater		**IGS** Variation in flow regimes with depth by determination of induced velocities in borehole water columns using various logging techniques	**BRMCE** Analysis of groundwater flow	**CWPU IGS** Model studies of regional groundwater flows as inputs to regional systems studies of groundwater and combined use development, conservation and management	**TWA** Transmissivity distribution by mathematical model
	WRC Radiochemical studies of flow, retention time and dispersion phenomena		**BRMCE** Unconfined-confined flow close to wells and rivers		
		BTLGG Deep quarrying and groundwater movement in Carboniferous Limestone of Bristol region	**NEWCE** Groundwater movement resulting from mineral abstraction		
			ICLGL Influence of reservoirs on groundwater flow		

TABLE 3 WATER OF THE LITHOSPHERE — (c) GROUNDWATER (*continued*)

Instruments and Techniques		Physical Hydrology and Experimental Work		Theoretical Studies and Data Analysis	
		Groundwater outflow; river, soil water, groundwater interactions			
IGS Methods of interstitial water extraction		**WWA** Evaluation of water supply and river management potential of Great Oolite aquifer and effects of limestone quarrying on ground-water sources			**CWPU** Determination of ground-water component of river discharge
				TWA Conceptual models of groundwater storage and total catchment runoff	**RDGAB** Agricultural conse-quences of extensive groundwater extraction in E. Anglia
					UCLGL Mathematical modelling in groundwater studies
		Water quality; tracers, dating, chemistry, pollution			
WRC IGS Heat pulse flow meter for low flow measure-ment in boreholes	**BRMGL** Radiocarbon dating of groundwater	**IGS WRC** Behaviour of leacheates beneath selected landfills by drilling and sampling	**IGS WRC** Behaviour of nitrate in unsaturated and saturated zones of aquifers	**WRC** Mathematical modelling of groundwater movement and pollutants	
IGS Geochemical methods for water analyses	**BTLGG** Limestone hydrology and water tracing	**IGS** Natural elemental geochemistry to inves-tigate mass transport in induced recharge studies	**NWWA** Age of water below landfill sites		**CWPU** Chemical quality of groundwater and influences on water resources development
IGS Development of downhole hydrogeo-chemical probes	**IGS** Stable isotope studies of natural groundwaters; dating techniques using ^{14}C and ^{3}H	**SWA WRC CWPU** Hydrogeology of zone of salt-water contamina-tion of the Chalk and feasibility of its removal	**BTLGG** Limestone hydrology and water tracing **NWWA** Geochemistry of Wirral Permo-Triassic sand-stone aquifer		

TABLE 3 WATER OF THE LITHOSPHERE — (c) GROUNDWATER *(continued)*

Instruments and Techniques	Physical Hydrology and Experimental Work		Theoretical Studies and Data Analysis
	IGS Geochemistry of UK groundwater – model and process studies, major and trace element occurrence	**IGS** Geochemistry of formation waters and brines	
	CRPB Groundwater quality beneath waste tips	**IGS** Geochemical studies of interstitial waters and host rocks in UK aquifers, especially the Chalk	
	Water quality; tracers, dating, chemistry, pollution		
		UCLGL Distribution of saline groundwaters in SE England	
	DGWE Survey of groundwater from waste tips using natural tritium as an indicator	**BRMGL** Trace element ground-water studies	
	SHFGL Geochemical investi-gation, east Derbyshire	**UCLGL** Distribution of saline groundwater, SE England	
	STRGG Chemistry of karst water	**SHFGL** Influence of Quaternary deposits on composition of groundwater in a Triassic aquifer	

TABLE 3 WATER OF THE LITHOSPHERE — (c) GROUNDWATER (*continued*)

Instruments and Techniques	Physical Hydrology and Experimental Work		Theoretical Studies and Data Analysis	
	Civil Engineering Aspects			
IGS Development of sectional lightweight shallow well linings and screens	**IGS** Groundwater studies in connection with underground oil storage, motorway drainage, new coalfields	**BRE** Influence of groundwater on stability of ground in connection with building work		
			Well inventories and maps	
			IGS Metric well inventories National and regional hydrogeological maps	Water supply contributions to IGS memoirs for one-inch geological maps

TABLE 4 DRAINAGE BASIN SYSTEMS — (a) GEOMORPHOLOGY AND RUNOFF PROCESSES

EANEV
Effects of water on slope recession

EXEGG
Drainage basin changes and their hydrological significance

CAMEN
River networks in south-west England

IH LDSGG
Contributing area and channel studies

CAMGG EXEGG UCLGG
Stream channel geometry

WABGG SHEGG
River channel processes

EXEGG
Flood plain and valley floor development in the Upper Exe basin

KCLGG
Mass movement on the Dorset coast

IH CAMEN
Initial development of stream channel networks

EANEV
Alluvial channels: processes and form

NEWCE OXFGG ICLCE
Channel morphology and runoff

SHEGG
Bed slope adjustment in natural channels

WABGG
Flood plain geometry

IH
Variation in catchment response

WWA
Mendip karst hydrology

IH EXEGG QMCGG
Influence of catchment characteristics on runoff

SHEGG
Effects of drainage basin morphology on flood characteristics

EANEV
Effects of drainage network geometry on stream hydrograph

IH
Storm runoff through natural pipes

LARS
Effect of weed-free environment, achieved by herbicides, on runoff

IH
Runoff from impervious surfaces

CAMGG IH
River discharge characteristics and associated basin forms

EANEV
Some effects of water on slope recession

OXFGG
Solution and percolation rates of selected limestones

MANGG
Role of an upland peat bog as a streamflow regulator in the hydrological system of a small catchment

SALCE
Slope, soil properties and rainfall intensity on infiltration and runoff

HULGG
Throughflow on hill-slopes, North Yorkshire moors

STRGG
Hillslope hydrology and water quality

EXEGG
Physical land classification in south-west England

EXEGG
Interrelationship of streamflow and quality in a small catchment

TABLE 4 DRAINAGE BASIN SYSTEMS — (b) EROSION AND SEDIMENT TRANSPORT

IH
Erosion and sediment transport in upland catchments

BRPCE
Mobilization and transport of sediment on chalk slopes

Sediment transport in Sussex streams

BTLGG
Suspended sediments in tributary rivers of Severn estuary

EDGFN
Erosion and sediment yield studies in southern Scotland

HULGG
Sediment production, Yorkshire Moors

STWA
Hydraulic and sediment studies, Hollinsclough experimental catchment; River Trent; reservoired area River Derwent; River Lin; Kingston catchment

BTLCE
Dispersion induced by density effects and sedimentation

EANEV
Water and suspended sediment movement in Breydon Water, East Anglia

EXEGG
Sediment and solute dynamics of Devon catchments

NEWCE
Pits for bed-load measurement

DUNGL
Bed load and suspended sediment budgets in Tay basin

HRS
Sediment transport relationships

Behaviour of rivers with gravel beds

STRCE
Mathematical models of erosion-transport-deposition process

DUNGL
Bed load and suspended sediment budget – Tay basin

UCLGG
Quantitative analysis of soil erodibility based on physical and chemical measurements

SHEGG
Mechanics of bank erosion

WABGG
Fluvial disperson of mining sediments

Solutes and sediments in mid-Wales

STRCE
Sediment bed load in parallel and converging channels

ABDGG
Hydraulic geometry, sediment transport, coastal morphology and pollution dispersion in Dee, Don and Ytham estuaries

HRS
Erosion and deposition characteristics of estuary muds

RDGGG
Sediment and chemical balance of two Welsh catchments

FBA
Suspended solids in relation to flow, chemical and bacterial content

NCAFE
Soil erosion in the UK

OXFGG
Clear water erosion: its geographical and international significance

TABLE 4 DRAINAGE BASIN SYSTEMS — (c) WATER BALANCE STUDIES

HULGG
Climate and water balance in a clay catchment

NEWGG
Effective rainfall reaching a channel in an upland catchment

EXEGG
Hydrology of Slapton Ley catchment area

BRDCE GLACE
Hydrology of small catchments

OXFAS
Effects of field drainage and of climatic change on three small catchments in eastern England

ASTCE
Computer simulation of the hydrologic cycle

PLPES
Integrated study of small upland granite catchment, S.W. Dartmoor

STAGG
Hydrological processes in Scottish upland catchment

STRGG
Thermal characteristics and related hydrology of upland catchments

WSWGG
Runoff and catchment characteristics in selected Breconshire catchments

IGS/IH
Permeable catchment study – field relationships between rainfall/infiltration/recharge/natural discharge in representative permeable outlier

STWA
Catchment and sediment studies of Blackbrook and Kingston catchments

Long-term rainfall deficits in Severn basin and relationship with groundwater levels

NIDA/DENI
Woodburn catchment, near Belfast

Quality/quantity model of River Main basin

CWPU/WAs
Dee (WNWDA/MO/WRC) River flow regulation study

Thet (AWA)
Development of groundwater for river regulation

NWWA
Arley Brook experimental catchment – water balance techniques and correlation between actual and potential evapotranspiration in terms of soil moisture profile

IH
Water balance of upland forested and grassed catchments, Plynlimon

Hydrograph prediction in River Ray catchment

IH/AWA
Water balance in the Cam catchment

IH/NWWA
Water balance and sediment studies, Coalburn catchment

SWA
Bartley mill catchment study

WWA
Climatological and water balance studies, Chew Stoke

WWA
Monitoring hydrological cycle within small drainage basin

BMNRPB
Comparative study of catchments containing natural lochs

NSHEB
Water balance and evaporation estimates for small catchment, Argyll

Catchment response to heavy rainfalls

ICLCE
Hydrology of laboratory catchments

HULGG
Catchwater catchment experiment

EXEGG
Application of grid square technique to hydrological mapping

LIVGG
Vegetation and hydrology of Ainsdale national sand dune nature reserve

TABLE 5 OPERATIONAL HYDROLOGY — (a) FLOOD HYDROLOGY

ICLCE
Analytical model for transient floods in presence of persistent base-flow

BTLCE
Attenuation of storm runoff by flood plains

MIPGG
Flood hazard research project

MO
Probable maximum precipitation (PMP) and statistical studies of heavy rainfall for selected catchments with plentiful data

EDGFN
Land use adjustment to flood risk, southern Scotland

BRMCE
Regional flood analysis based on stochastic correlation and physical parameters

SWWA
Investigations into 1968 floods in the south-west

EANEV
Flood alleviation feasibility

LANEV
Flood design by unit hydrograph technique

BRMCE HRS
Application of flood routing methods to British rivers

NEWCE
Snowmelt river flooding in UK

STWA NOTCE IH
River flood frequencies

IH
Flood estimation for gauged and ungauged catchments

NSHEB
Flood routing in hydro-electric scheme cascades

WABGG
Floodplain geometry

TABLE 5 OPERATIONAL HYDROLOGY — (b) MATHEMATICAL MODELS FOR HYDROLOGY

ASTCE
Mathematical models of precipitation, evaporation and snowmelt runoff in mountainous regions

NEWCE
Simulation by conceptual modelling

ICLCE
Optimisation of conceptual catchment models

NEWCE
Unit response functions

ICLCE
Small sample properties of stationary stochastic models in hydrology

STRCE
Developing and testing mathematical models in hydrology

ICLCE
Analysis of hydrological time series

KCLCE
Modelling hydrologic systems by finite element method

ICLCE
Conceptual models of rainfall-runoff process

NEWCE
Stochastic-deterministic models

SOTCE
Extension of rainfall and runoff records using multiple regressions approach

WSWGG
Digital simulation of drainage basins

ICLCE
Models for synthetic data generation

WNWDA
Catchment models based on ISO functions

SWWA
Monthly rainfall-runoff modelling

SALCE
Simulation of catchment response

ICLCE
Time-variant catchment models

EANEV
Deterministic, spatially distributed model of catchment hydrology

CWPU
Linear and semi-logarithmic catchment regression models

IH
Techniques for predicting catchment response to rainfall

IH EXEGG
Study of regional low flow characteristics relating basin characteristics to seasonal low flow regimes

NCAFE
Prediction of runoff from small catchments

CWPU BNMCE
Critical review of use of multiple regression analysis methods

SWWA
Conceptual and time series mathematical models

LANEV
Introduction of stochastic elements into deterministic catchment modelling

Non-linear conceptual catchment model for isolated storm events

CWPU
Accuracy of catchment models related to catchment characteristics

TABLE 5 OPERATIONAL HYDROLOGY — (c) WATER RESOURCES MANAGEMENT

AWA
Hydrological operation of Ely Ouse-Essex interbasin transfer scheme

BRMCE
Derivation of optimal control rules for water resource systems

CTYCE
Computer simulation of water resource systems

EXEGG
Evaluation and mapping of water resource characteristics of Dartmoor

MO
Assessment of areal rainfall, evaporation and soil moisture deficit for water resources studies

NWA
Development of real-time operating system for Kielder scheme; optimisation studies of water resource systems applied to Kielder scheme

NWWA
Operation of multi-source water supply schemes (especially Lancashire conjunctive use scheme)

LANEV
Optimal design of water resources systems

ICLCE
Decision-making in design of water resource systems

SNPCE
Analysis of river regulating scheme

WRC
Studies to link surface and groundwater resources, desalination plants and industrial reclamation

WISCE
Water economy of Great Britain

WRC
Operational research on complex resource systems, e.g. interconnected reservoirs, pumped storage, river regulation scheme

CWPU
Hydrological control of optimal combined use of surface water resources

IH CWPU WRC
Development of flow forecasting model for Dee

WRC
Time series analysis to interpret quality, quantity and flow variations

TWA
Operating rules for injunctive use of resources

STWA
Operational models for flow forecasting on regulated River Severn

CWPU WRC
Flow forecasting using radar measurement of rainfall over Dee

SURCE
Analysis of river regulating system

SWA
Optimization of conjunctive use of water resources

CWPU
Review of suspect stream flows and effects on water resource planning

CWPU WAs
Groundwater assessment, development and conservation

BQUGG
River flood and water supply problems in Northern Ireland

CWPU WAs
Groundwater resources of selected UK aquifers

HRS
Economics of flood alleviation

SWA/WRB
Computer systems for archiving and processing water balance data

CWPU
Relationship between surface and groundwater and influences on water resource development

CWPU
Water resources maps

STWA
Craig Goch and Dover Beck schemes

TABLE 5 OPERATIONAL HYDROLOGY — (d) URBAN HYDROLOGY AND EFFLUENT PROBLEMS

DGWE
Design and performance of sewerage systems; relationship between rainfall and sewer flows

IH
Prediction of runoff from partially or totally urbanised areas

SHECE
Statistical treatment of urban drainage

TRRL
Storm water sewer design; bridge and culvert design

BRMCE
Runoff from urban catchments via surface channels and stormwater sewers

ICLCE UCLGG
Hydrological consequences of urbanisation

SHECE
Urban rainfall and runoff modelling

MO
Effects of travelling storms on urban drainage

BTLCE EXEGG ICLCE UCLGG STWA
Effect of urban development on catchment hydrology

STWA
Design criteria for storm stations on trunk sewers

EXEGG
Building activity and urbanisation on streamflow, sediment and solute yield

LANEV
Water chemistry and sediment in motorway drainage water

ICLGL
Underground disposal of waste in the UK

STWA
Treatment of sewage with metallic salts and industrial wastes

DENI
Drainage and effluent flow regimes from new housing estates

NWWA
Paper mill effluent

STWA
Analysis of motorway runoff

STWA
Impact on fishing of effluent input

CHAPTER III

HYDROLOGICAL RESEARCH BY GOVERNMENT AND OTHER AGENCIES

The study of hydrology in Britain has developed slowly and on an *ad hoc* basis; nevertheless much useful research has been done in certain fields such as precipitation, evaporation and groundwater, usually incidental to a research programme such as land drainage.

The diverse origins of hydrological research in the United Kingdom are reflected in the number and character of the agencies listed below. Interests range from the data accumulation and monitoring activities of various river and water authorities through applied project studies by such organisations as the Transport and Road Research Laboratory to the more fundamental research of the Institute of Hydrology.

The purpose of this chapter is to describe briefly the interest in hydrology of each agency. At the end of each entry is a selected bibliography of publications prepared since 1968.

DEPARTMENT OF EDUCATION AND SCIENCE

AGRICULTURAL RESEARCH COUNCIL

EAST MALLING RESEARCH STATION
East Malling, Maidstone, Kent

East Malling Research Station, grant aided by the Agricultural Research Council, devotes all its resources to studying the culture of deciduous fruits and the pests and diseases of hops. Many of the experiments are undertaken to obtain immediate practical answers to the problems of fruit growers, but some are of a more fundamental nature and contribute to better understanding of the growth and behaviour of the fruit plant.

The work of the Research Station is shared between nine closely related sections, one of which (the Fruit Nutrition Section) is responsible for studies on soil-water-plant relationships as well as on crop nutrition.

Soil moisture measurements are made in the field using a neutron moisture meter and tensiometers installed at different depths through the profile.

Water stress is measured in the tree and its effects, together with those of its components, osmotic and turgor pressures, are studied in relation to tissue contraction and stomatal closure in the shorter term and growth and cropping of the trees in the longer term.

Lines of study are aimed at defining tolerable limits of plant water stress by identifying 'true' water stress (in the physiological sense). For this purpose a metabolic change in the plant is being used as the index of the commencement of stress.

Nutritional studies in relation to the soil water supply are also being made.

Goode, J. E. 1970. Cumulative effects of irrigation on fruit crops. *Proc. Symp. XII on The role of water in agriculture*. Univ. Coll. of Wales, Aberystwyth, Pergamon Press, Oxford.

Goode, J. E. and Hyrycz, K. J. 1970. The response of black currants to different soil moisture conditions and two levels of nitrogenous fertilizer. *J. hort. Sci., 45,* 379-91.

Goode, J. E. and Ingram, J. 1971. The effect of irrigation on the growth, cropping and nutrition of Cox's Orange Pippin apple trees. *J. hort. Sci., 46,* 195-208.

Goode, J. E. and Higgs, K. H. 1973. Water, osmotic and pressure potential relationships in apple leaves. *J. hort. Sci., 48,* 203-15.

Goode, J. E. 1974. Water relations and nutrition of trees. In: *Tree Growth in the landscape,* Wye College, Kent, 5-9.

Pereira, H. C. (Editor) 1975. *Climate and the orchard.* Research Review No. 5, Commonwealth Bureau of Horticulture and Plantation Crops, East Malling, Maidstone, Kent.

GRASSLAND RESEARCH INSTITUTE
Hurley, Maidenhead, Berkshire SL6 5LR

The Grassland Research Institute, grant aided by the Council, is concerned with the growth and utilization of forage crops that make the maximum use of the radiant energy of the sun and the nutrients available in the soil.

The Agronomy Department undertakes research into irrigation and the efficiency of water use by grass and forage crops. Particular attention is given to the rate of uptake of soil water by different crops and from different soils and the effects of differences in water uptake on crop growth rate. The relationship between the location of water in the soil profile and the availability of plant nutrients is considered. Work is carried out in collaboration with the Botany Department on the use of chemical desiccants applied before cutting to decrease the water content of the crop at harvest and on the loss of water from forage cut for conservation.

The Botany Department undertakes measurement of radiation, temperature, relative humidity, wind speed and carbon dioxide content within and above the herbage crop and makes direct measurements of carbon dioxide and water vapour exchange of the crop. The Department is concerned particularly with the relationship between the aerial environment, leaf diffusion resistance and leaf water status of grasses and with the relationship between crop transpiration, carbon dioxide exchange and crop growth. Another interest is the loss of water from cut forage crops during field drying and the relationship between the swath microclimate and the prevailing weather.

Brockington, N. R. 1970. Herbage Growth. In: Jones, J. G. W. (Ed.). The use of models in agricultural and biological research. Hurley, The Grassland Research Institute, 16-24.

Clement, C. R., Jones, L.H.P. and Hopper, M. J. 1972. The leaching of some elements from herbage plants by simulated rain. *Jnl. appl. Ecol.,* 9, 249-60.

Garwood, E. A. 1969. Seasonal tiller populations of grass and grass/clover swards with and without irrigation. *J. Br. Grassld Soc., 24,* 333-44.

Garwood, E. A. 1968. Some effects of soil-water conditions and soil temperature on the roots of grasses and clover. 2. Effect of variation in soil-water content and soil temperature on root growth. *J. Br. Grassld Soc., 23,* 117-28.

Garwood, E. A. 1974. Use of water by grassland and response to partial irrigation. *12th int. Grassld Congr. Moscow.*

Garwood, E. A. and Tyson, K. C. 1973. Losses of nitrogen and other plant nutrients to drainage from soil under grass. *J. agric. Sci., Camb., 80,* 303-12.

Garwood, E. A. and Tyson, K. C. 1973. The response of S24 perennial ryegrass swards to irrigation. I. Effects of partial irrigation on DM yield and on the utilization of applied nitrogen. *J. Br. Grassld Soc., 28,* 223-33.

Garwood, E. A. and Tyson, K. C. 1975. The response of S24 perennial ryegrass to irrigation. II. Variation in soil- and plant-water status. *J. Br. Grassld Soc., 30,* 51-62.

Garwood, E. A. and Williams, T. E. 1967. Soil water use and growth of a grass sward. *J. agric. Sci., 68,* 281-292.

Garwood, E. A. and Williams, T. E. 1967. Growth water use and nutrient uptake from the subsoil by grass swards. *J. agric. Sci., 69,* 125-130.

Harris, C. E., Thaine, R. and Sarisalo H. I. Marjatta. 1974. Effectiveness of some mechanical, thermal and chemical laboratory treatments on the drying rates of leaves and stem internodes of grass. *J. agric. Sci., Camb., 83,* 353-8.

Leshem, Y., Thaine, R., Harris, C. E. and Canaway, R. J. 1972. Water loss from cut grass with special reference to hay making. *Ann. appl. Biol., 72,* 89-104.

Sheehy, J. E., Green, R. M. and Robson, M. J. 1975. The influence of water stress on the photosynthesis of a simulated sward of perennial ryegrass. *Ann. Bot., 39,* 387-401.

Stiles, W. 1970. A diffusive resistance porometer for field use. I. Construction. *Jnl. appl. Ecol., 7,* 617-22.

LETCOMBE LABORATORY
Wantage OX12 9JT

The Letcombe Laboratory is concerned with the growth of crops in relation to soil conditions, including drainage, and the long term influence of different methods of cultivation on soil structure and fertility. Detailed laboratory studies and field experiments are closely co-ordinated.

Work of hydrological interest arises mainly from work of the Drainage Section studying the influence of transient waterlogging on crop growth. Lysimeters containing undisturbed columns of soil are installed in a substructure so that their surfaces are level with the surrounding soil which carries a guard crop. A mobile glasshouse with a traversing irrigation system allows a predetermined rainfall pattern to be simulated. Tunnels under the guard crop give access to the lysimeters for regulation of the water-tables, collection of drainage water, sampling the soil water and atmosphere, and monitoring soil temperature. Variations in moisture profiles are measured with a neutron meter. Nitrogen-15 is being used as a tracer to study the utilization of nitrogen fertilizer by the crop, and leaching through soil.

In the laboratory, studies are in progress on the size and continuity of soil pores in relation to the movement of both water and gases. A small investigation is in progress into the relative movement of tritium-labelled water and nitrate down columns of undisturbed chalk; this is relevant to studies elsewhere of the leaching of fertilizers into aquifers.

The Electronics Section have adapted a cheap cassette tape recorder for use with a standard tipping bucket raingauge. It will run unattended for up to six weeks and provides hourly records of rainfall.

Agricultural Research Council Letcombe Laboratory Annual Report, 1973, 1974. Agricultural Research Council Letcombe Laboratory Annual Report 1974, 1975.

LONG ASHTON RESEARCH STATION
Long Ashton, Bristol BS18 9AF

Long Ashton Research Station is grant aided by the Council.

Hydrological interests, recently developed, are mainly concerned with three aspects, namely, (a) the effects of altitude, slope and aspect on microclimate, including evaporation rates, (b) the fate of persistent insecticides, and their movement within and through the soil into drainage, and (c) the effects of a weed-free environment, achieved by herbicides, on soil cover, infiltration and runoff (in initial stage of work).

Hudson, J. P. 1968. Lysimeters and weighable containers in greenhouse experimentation. *Symposium on Techniques of Experimentation in Greenhouses. Tech. Comm. 7 of I.S.H.S.*, 104-114.

Hudson, J. P. 1968. The plant and its environment: I. Effects of weather on plants. II. Effects of plants on weather. (48th Masters Memorial Lecture, 1967). *J. roy. Hort. Soc., XCIII*, 193-209 and 243-254.

Hudson, J. P. 1973. Water relations and irrigation of temperate fruits. *Proc. 18th Int. hort. Congress, Tel Aviv, Israel, 1970, 4*, 199-210.

Waister, P. D. and Hudson, J.P. 1970. Effects of soil moisture regimes on leaf water deficits, transpiration and yields of tomatoes. *J. hort. Sci., 45*, 359-370.

MACAULAY INSTITUTE FOR SOIL RESEARCH
Craigiebuckler, Aberdeen

The Macaulay Institute for Soil Research is grant aided by the Department of Agriculture and Fisheries for Scotland on the advice of the Agricultural Research Council. The research carried out is designed to obtain information on the fundamental properties of soil and soil-plant relationships, the object being to maintain and improve soil fertility.

Investigations into soil moisture characteristics are undertaken in relation to plant growth, particularly on the highly organic soils that cover considerable areas in northern and western Scotland. Work on the moisture status and water movement in peat with respect to forest growth is in progress using various techniques including tritium-labelled water. The chemical composition of precipitation is being assessed with regard to nutrient input, and measurement of interception and penetration in forest stands is being carried out in studies of nutrient movement in such ecosystems. In relation to the drainage classes of soils as defined by the Soil Survey of Scotland, water table levels in profiles of a hydrological sequence of soils on granitic till have been recorded at weekly intervals for four years.

In work on the characterization of the physical properties of soil series, the programme includes measurements of relevant parameters such as moisture release characteristics, water storage capacity, drainable porosity and hydraulic conductivity.

Three maps and explanatory pamphlets entitled 'Assessment of Climatic Conditions in Scotland' have been published.

Birse, E. L. and Dry, F. T. 1970. Assessment of climatic conditions in Scotland. 1. Based on accumulated temperature and potential water deficit. (Map and Explanatory Pamphlet). 25pp. Aberdeen: Macaulay Institute for Soil Research.

Birse, E. L. and Robertson, L. 1970. Assessment of climatic conditions in Scotland. 2. Based on exposure and accumulated frost. (Map and explanatory Pamphlet). 41pp. Aberdeen: Macaulay Institute for Soil Research.

Birse, E. L. 1971. Assessment of climatic conditions in Scotland. 3. The bioclimatic sub-regions. (Map and Explanatory Pamphlet). 12pp. Aberdeen: Macaulay Institute for Soil Research.

Boggie, R. 1972. Effect of water-table height on root development of *Pinus contorta* on deep peat in Scotland. *Oikos, 23*, 304-312.

Knight A. H., Boggie, R. and Shepherd, H. 1972. The effect of ground water level on water movement in peat: a study using tritiated water. *J. appl. Ecol.*, 9, 633-641.

NATIONAL VEGETABLE RESEARCH STATION
Wellesbourne, Warwickshire

National Vegetable Research Station is mainly concerned with research into methods of improving the yield and quality of vegetables.

Recent hydrological work has been concerned with the water balance of cropped and uncropped soil, as it affects the movement and uptake by plants of nutrients and herbicides in the soil.

A means of forecasting the distribution of nitrogen in soil and its availability to plant roots from standard meteorological data and laboratory measurements on soil is being developed. The work involves providing experimental evidence to test a model for upward and downward movement of ions, studying the effect of soil moisture and texture on the transport of nitrate through the soil to roots, and developing quantitative information about microbial activities.

A computer model for the redistribution of water in fallow soil has been developed and is being tested in field and laboratory experiments. The model is being extended to account for the uptake of water by plants using standard meteorological data and laboratory measurement of soil properties. The work includes the construction of simple empirical models for the spatial development of root systems in the soil.

Quantitative relations about soil/plant processes are being combined into dynamic models for predicting the effects of various agronomic practices, and of soil and weather conditions on the responses of crops to fertilizers in field. The validity of the models are being tested in detailed experiments at Wellesbourne, at EHS Stations and on growers' holdings.

A model has been developed for predicting the

persistence of soil-applied herbicides from laboratory measurements and standard meteorological data. The models are being tested in both field and glasshouse experiments.

Barnes, A., Greenwood, D. J. and Cleaver, T. J. (In press). A dynamic model for the effects of potassium and nitrogen fertilizers on the growth and nutrient uptake of crops. *J. Agric. Sci., Camb.*

Burns, I. G. 1974. Leaching of plant nutrients through soils. *Reps. on Prog. of App. Chem., 58,* 408-20.

Burns, I. G. 1974. A model for predicting the redistribution of salts applied to fallow soils after excess rainfall or evaporation. *J. Soil Sci., 25,* 165-78.

Burns, I. G. 1974. A simple method of predicting nitrate leaching in the field. *J. Sci. Food Agric., 25,* 1196-7.

Gerwitz, A. and Page, E. R. 1974. Empirical mathematical model to describe plant root systems. *J. app. Ecol., 11,* 773-82.

Greenwood, D. J., Cleaver, T. J. and Niendorf, K. B. 1974. Effects of weather conditions on the response of lettuce to applied fertilizers. *J. Agric. Sci., Camb., 82,* 217-32.

Greenwood, D. J., Wood, J. T. and Cleaver, T. J. 1974. A dynamic model for the effects of soil and weather conditions on nitrogen response. *J. Agric. Sci., Camb., 82,* 455-67.

Greenwood, D. J., Wood., J. T., Cleaver, T. J. and Hunt, J. 1971. A theory for fertilizer response. *J. Agric. Sci., Camb., 77,* 511-23.

Page, E. R. (In press). The location and persistence of ammonia (aqueous, anhydrous and anhydrous + 'N serve') injected into a sandy loam soil, as shown by changes in concentration of ammonium and nitrate ions. *J. Agric. Sci., Camb., 87.*

Page, E. R. and Gerwitz, A. 1974. Mathematical models based on diffusion equations, to describe root systems of isolated plants, row crops and swards. *Plant and Soil, 41,* 243-54.

Rowse, H. R. (In press). Simulation of the water balance of soil columns and fallow soils. *J. Soil Sci., 26.*

Walker, A. (In press). A simulation model for prediction of herbicide persistence. *J. Environ. Qual., 4.*

ROTHAMSTED EXPERIMENTAL STATION
Harpenden, Hertfordshire

Rothamsted Experimental Station, grant aided by the Agricultural Research Council, is the largest and oldest of agricultural research centres. Most of the departments are biological, being concerned with the plant in health and disease. Of the non-biological sections, the Physics Department divides its interest between the root environment (soil physics) and the leaf environment (agricultural meteorology).

Studies of soil water come into both spheres of interest and there has been assembled a considerable body of information and a number of ideas on the relationship between the use of water by plants and the crop environment. Irrigation experiments are showing the dependence of crop yields on potential evaporation.

French, B. K., Long, I. F. and Penman, H. L. 1973. Water use by farm crops.
I Test of the neutron meter on barley, beans and sugar beet 1970.
II Spring wheat, barley, potatoes 1969; Potatoes, beans, kale 1968.
III Bare soil, short turf and crops in rotation, 1962 to 1967, 1971.
Rep. Rothamsted exp. Sta. for *1972, Part 2*, 5-42, 43-61, 62-85.

Legg, B. J. and Long, I. F. 1973. Microclimate factors affecting evaporation and transpiration. In: *Ecological Studies. Analysis and Synthesis, Vol. 4* (Ed. A. Hadas *et al*), Springer Verlag.

Monteith, J. L. and Bull, T. A. 1970. A diffusive resistance porometer for field use.
II Theory, calibration and performance.
J. appl. Ecol., 7, 623-638.

Parkinson, K. J. and Legg, B. J. 1972. A continuous flow porometer. *J. appl. Ecol., 9*, 669-675.

Penman, H. L. 1969. Water as a factor in productivity. In: *Proc. Symp. on Potential Crop Production in Britain*, Aberystwyth.

Penman, H. L. 1970. Woburn Irrigation 1960-68.
IV. Design and interpretation.
V. Results for leys.
IV. Results for rotation crops.
J. Agric. Sci., Camb., 75, 69-73, 75-88, 89-102.

Penman, H. L. 1970. The water cycle. *Scient. Am. 223*, 99-108.

Penman, H. L. 1971. Irrigation at Woburn VII. *Rep. Rothamsted exp. Sta. for 1970, Part 2*, 147-170.

Prebble, R. E. and Currie, J. A. 1970. Soil water measurement by a low resolution nuclear magnetic resonance technique. *J. Soil Sci., 21*, 273-288.

Rose, D. A. 1971. Water movement in dry soils: II An analysis of hysteresis. *J. Soil Sci., 22*, 490-507.

Rose, D. A. 1973. Some aspects of hydrodynamic dispersion of solutes in porous materials. *J. Soil Sci., 24*, 284-295.

Stiles, W. 1970. A diffusive resistance porometer for field use. I. Construction. *J. Appl. Ecol., 7*, 617-622.

SCOTTISH INSTITUTE OF AGRICULTURAL ENGINEERING

Bush Estate, Penicuik, Midlothian EH26 0PH

The Scottish Institute of Agricultural Engineering is grant aided by the Department of Agriculture and Fisheries for Scotland on the advice of the Agricultural Research Council. The Institute has a small research team studying the effects on soil physical conditions of cultivations and traffic by agricultural machinery, with an emphasis on understanding and reducing the damage to soil structure caused by some modern farming practices.

Part of this work is concerned with monitoring soil moisture regimes and soil hydraulic properties in long-term cultivation experiments in Scotland, in which the cultivation techniques range from traditional deep ploughing through some of the more recent reduced cultivation techniques to zero tillage. The effects of these practices on the movement and retention of soil water are being assessed.

Soane, B. D., Campbell, D. J. and Herkes, S. M. 1973. Cultivation techniques in relation to continuous barley growing. II Soil Physical Conditions. *Proc. Int. Conf. Res. Methods* (*Silsoe, 1970*) 58-76.

Soane, B. D. and Pidgeon, J. D. 1975. Tillage requirements in relation to soil physical properties. *Soil Science, 119* (*5*), 376-84.

SOIL SURVEY OF ENGLAND AND WALES
Rothamsted Experimental Station, Harpenden, Hertfordshire

The Soil Survey of England and Wales is grant aided by the Agricultural Research Council and exists to organise information about soil variability in classifications, maps and explanatory texts. These are used to transfer results of research and experience in crop production and soil management from one tract of land to another, in planning land use, and in planning further research in soil science or allied fields.

Experience has shown that properties used to identify and map soils reflect infiltration rate and moisture storage capacity, and are relevant to investigations of the pathways of water movement. The Survey is involved in field and laboratory measurements of water levels, hydraulic conductivity, bulk density, moisture release characteristics and porosity, to characterise areally important soil series in hydrological terms and to increase the reliability of inferences derived from properties assessed in the field. Inferences and measurements provide relevant data for catchment research and studies of the component processes of the hydrological cycle.

Mackney, D., Smith, P. D., and Thomasson, A. J. 1975. Winter rain acceptance potential: map 1:625,000 of Great Britain. In: *Flood Studies Report, Volume 5.*

Mackney, D. and Thomasson, A. J. 1975. Types of soil water regime; map 1:2,000,000 England and Wales. In: *Soils and Field Drainage. Tech. Monogr. No. 7.*

Jarvis, M. G. and Walpole, R. A. 1975. Predicted underdrainage treatment for arable land use: map 1:63,360. In: *Soils and Field Drainage, etc.*

Thomasson, A. J. (Ed.). 1975. Soils and Field Drainage Tech. Mongr. No. 7.

Rudeforth, C. C. and Thomasson, A. J. 1970. Hydrological properties of soils in the River Dee Catchment. *Spec. Surv. Soil Surv. No. 4.*

Reeve, M. J., Smith, P. D. and Thomasson, A. J. 1973. The effect of density on water retention properties of field soils. *J. Soil Sci., 24,* 355-367.

Smith, P. D. and Thomasson, A. J. 1974. Density and water-release characteristics. In: Soil Survey Laboratory Methods. *Soil Surv. Tech. Mongr.* No. 6, 42-56.

Thomasson, A. J. 1973. Factors influencing the water regimes of gleyed clayey soils in moist temperate regions. In: Pseudogley and gley. Ed. E. Schlichting and U. Schwertmann. Trans. Comm. V and VI *Inst. Soil Sci. Soc. Verlag Chemie*, 491-502.

Beckett, P. H. T. and Webster, R. 1972. Matric suction to which soils in south-central England drain. *J. agric. Sci., 78,* 379-387.

UNIT OF SOIL PHYSICS

219c Huntingdon Road, Cambridge, CB3 0DL

The work of the Unit developed from a consideration of land-drainage problems and now includes a study of the whole field of soil-water movement. The Unit studies the theory of the movement of water both in the insaturated zone above the water table and in the saturated groundwater zone. The former is of particular application to soil water profile development near the soil surface, while the latter gives a theoretical basis for groundwater movement and control which is relevant not only to field drainage design but to all groundwater flows.

Bybordi, M. 1969. The vertical movement of water in stratified porous material: 2. Transient stages of draining to a water table. *Wat. Resour. Res., 5,* 694-697.

Childs, E. C. 1969. *An introduction to the physical basis of soil water phenomena.* John Wiley and Sons, Ltd., London. 493pp.

Childs, E. C. 1971. Drainage of groundwater resting on a sloping bed. *Wat. Resour. Res., 7,* 1256-1263.

Childs, E. C. 1972. Concepts of soil water phenomena. *Soil Sci.,113,* 246-253.

Childs, E. C. and Bybordi, M. 1969. The vertical movement of water in stratified porous material: 1. Infiltration. *Wat. Resour. Res., 5,* 446-459.

Childs, E. C. and Tzimas, E. 1971. Darcy's law at small potential gradients. *J. Soil Sci.,* 22, 319-327.

Childs, E. C. and Youngs, E. G. 1974. Soil Physics: Twenty-five years on. *J. Soil Sci., 25,* 420-428.

Gureghian, A. B. and Youngs, E. G. (In press). The calculation of steady-state water-table heights in drained soils by means of the finite element method. *J. Hydrology.*

Marei, S. 1974. A Hele-Shaw model study of oscillating water tables in drained homogenous soils. *Soil Sci., 117,* 301-305.

Marei, S. M. and Towner, G. D. 1975. A Hele-Shaw analogue study of the seepage of groundwater resting on a sloping bed. *Wat. Resour. Res.*

Poulovassilis, A. 1969. Steady-state potential and moisture profiles in layered porous media. *Soil Sci., 107,* 47-52.

Poulovassilis, A. 1972. The changeability of the hydraulic conductivity of saturated soil samples. *Soil Sci., 113,* 81-87.

Poulovassilis, A. 1973. A method for studying soil-water properties of soil profiles. *Soil Sci., 24,* 117-124.

Poulovassilis, A. 1974. The uniqueness of the moisture characteristics. *Soil Sci., 25,* 27-33.

Poulovassilis, A. and Childs, E. C. 1972. The hysteresis of pore water: the non-independence of domains. *Soil Sci., 112,* 301-312.

Poulovassilis, A., Krentos, V. D., Stylianou, Y. and Metochis, Ch. 1974. Soil-water properties of a layered soil determined *in situ*. *Proc. Symp. Isotope and Radiation Techniques in Soil Physics and Irrigation Studies;* IAEA/FAO, Vienna, 1973. 205-223.

Poulovassilis, A. and Tzimas, E. 1974. The hysteresis in the relationship between hydraulic conductivity and suction. *Soil Sci., 117,* 250-256.

Santos Junior, A. C. dos and Youngs, E. G. 1969. A study of the specific yield in land-drainage situations. *J. Hydrology, 8,* 59-81.

Swartzendruber, D. and Youngs, E. G. 1974. A comparison of physically-based infiltration equations. *Soil Sci., 117,* 165-167.

Towner, G. D. 1975. Drainage of groundwater resting on a sloping bed with uniform recharge. *Wat. Resour. Res., 11,* 144-147.

Youngs, E. G. 1969. Unconfined aquifers and the concept of the specific yield. *Bull. int. Ass. Sci. Hydrol., XIV* (2), 191-197.

Youngs, E. G. 1970. Hodograph solution of the drainage problem with very small drain diameter. *Wat. Resour. Res., 6,* 594-600.

Youngs, E. G. 1970. Some aspects of water movement through soils. In: *S. C. I. Monograph No. 37. Sorption and transport processes in soils,* 107-119.

Youngs, E. G. 1971. Optimum pumping conditions for wells located in unconfined coastal aquifers. *J. Hydrol, 13,* 63-69.

Youngs, E. G. 1971. Seepage through unconfined aquifers with lower boundaries of any shape. *Wat. Resour. Res., 7,* 624-631.

Youngs, E. G. 1971. Steady state flow around wells in aquifers with hydraulic conductivity varying with depth. *Wat. Resour. Res., 7,* 1366-1368.

Youngs, E. G. 1972. Two- and three-dimensional infiltration seepage from irrigation channels and infiltrometer rings. *J. Hydrol., 15,* 301-315.

Youngs, E. G. 1974. Seepage rates and the horizontal flow approximation. *Wat. Resour. Res., 10,* 874-876.

Youngs, E. G. 1974. The use of the Green and Ampt approach in soil-water movement studies. *Proc. Xth Int. Symp. Agrochimica, Water in Agriculture,* Bari, Italy. *Agrochimica XVIII,* 231-239.

Youngs, E. G. 1974. Water-table heights in homogenous soils drained by non-ideal drains. *Soil Sci., 117,* 295-300.

Youngs, E. G. 1975. The effect of the depth of an impermeable barrier on water-table heights in drained homogenous soils. *J. Hydrol., 24,* 283-290.

Youngs, E. G. and Towner, G. D. 1970. Comments on 'Hydrostatics and hydrodynamics in swelling soils' by J. R. Philip. *Wat. Resour. Res., 6,* 1246-1247.

WELSH PLANT BREEDING STATION
Plas Gogerddan, Aberystwyth, Dyfed

The Welsh Plant Breeding Station, grant-aided by the Agricultural Research Council, is concerned with the breeding of herbage grasses, legumes and arable crop varieties.

The Agronomy Department (Dr. H. Thomas, Mr. J. M. M. Munro) is studying the response of the grass sward to the physical environment and evaluating the physiological efficiencies of contrasting varieties. This work includes the measurement and control of water stress and the description of the effectiveness of differing water use strategies. Sward productivity and survival are related to climatic data, continuously recorded at a number of meterological sites, forming an altitudinal transect across mid-Wales. At Pant-y-dŵr

Hill Centre, lysimeters are used to compare leaching and water losses on contrasting soils under native and reseeded vegetation.

In the Department of Developmental Genetics (Dr. D. Wilson) the soil water use characteristics of contrasting grass varieties and lines selected for features influencing water transfer are being examined in small plots outdoors, using a Wallingford neutron soil moisture probe.

Thomas, Henry. 1975. The growth responses to weather of simulated vegetative swards of a single genotype of *Lolium perenne*. *J. agric. Sci., Camb.*, *84*, 333-343.

Munro, J. M. M. 1973. Potential pasture production in the uplands of Wales. 1. Climatic variation. *J. Br. Grassld Soc.*, *28*, 59-67.

Munro, J. M. M. and Davies, D. A. 1973. Potential pasture production in the uplands of Wales. 2. Climatic limitations on production. *J. Br.Grassld. Soc.*, *28*, 161-9.

Munro, J. M. M., Davies, D. A. and Morgan T. E. H. 1974. Research at Pant-y-dŵr Hill Centre. *Rep. Welsh Pl. Breed Stn for 1972*, 209-228.

Wilson, D. 1975. Leaf growth, stomatal diffusion resistance and photosynthesis during droughting of *Lolium perenne* populations selected for contrasting stomatal length and frequency. *Ann. appl. Biol.*, *79*, 67-82.

Wilson, D. 1975. Stomatal diffusion resistances and leaf growth during droughting of *Lolium perenne* plants selected for contrasting epidermal ridging. *Ann. appl. Biol.*, *79*, 83-94.

NATURAL ENVIRONMENT RESEARCH COUNCIL

BRITISH ANTARCTIC SURVEY

2 All Saints Passage, Cambridge, CB2 3LS

The British Antarctic Survey is responsible for initiating and executing scientific research in British Antarctic Territory (BAT) and the Falkland Islands Dependencies. Research of hydrological interest is largely carried out by the Glaciology section which is based at the Scott Polar Research Institute. This work falls into four categories:

1. An inventory of contemporary snow and ice in BAT is built up from studies of the areal extent of ice revealed in satellite and high flying aircraft photographs. This information is converted into ice volumes using ice depths determined by airborne radio echo sounding over large areas. Long-term changes are revealed by repeated measurements of reference surface height profiles of representative glaciers. The distribution and variability of sea ice from year to year is recorded and analysed with the help of satellite pictures and ship observations.
2. Two glaciers which represent the polar and sub-polar environments have been selected for detailed studies of ice, heat, and water balance. This work is designed to shed light on the relationship between glaciers and climate and to isolate those factors most responsible for changes in ice volume.
3. The study of ice deformation in floating ice shelves leads to an understanding of the stability of ice shelves, calving from which represents the single most important source of ice loss from the Antarctic

ice sheet. Bottom melting from the shelves is an important source of Antarctic Bottom Water.

4. Studies of the chemical properties of Antarctic snow and its included impurities give information about
   (a) chemical species which are related to climatic parameters and which may be sought in ice cores covering the past 1000 years in order to build up a picture of climatic changes during this period.
   (b) the global atmospheric circulation of pollutants and the natural baseline levels of these materials in snow deposited prior to the industrial revolution.

Coslett, P. H., Guyatt, M. and Thomas, R. H. 1975. Optical levelling across an Antarctic ice shelf. *BAS Bulletin*, No. 40, 55-63.

Doake, C. S. M. (In press). Glacier sliding measured by a radio echo technique. *J. Glaciol., 15* (73).

Peel, D. A. 1975. Organochlorine residues in Antarctic snow. *Nature, 234* (5498), 324-5.

Peel, D. A. (In press). Snow accumulation, conductance, and temperature inland from Halley Bay. *BAS Bulletin*.

Smith, B. M. Ewen. 1972. Automatic plotting of airborne geophysical survey data. *J. Navig., 25* (2), 162-175.

Smith, B. M. Ewen. 1972. Airborne radio echo sounding of glaciers in the Antarctic Peninsula. *BAS Sci. Rep., 72*.

Smith, B. M. Ewen and Evans, S. 1972. Radio echo sounding: absorption and scattering by water inclusion and ice lenses. *J. Glaciol., 11* (61), 133-146.

Swithinbank, C. W. M. 1970. Ice movement in the McMurdo Sound area of Antarctica. *Int. Symp. Antarctic Glaciological Exploration, Hanover, New Hampshire, U.S.A., 1968.* (Publ. 86 of Int. Ass. scient. *Hydrol.*), 97-115.

Swithinbank, C. W. M. 1970. Satellite photographs of the Antarctic Peninsula area. *Polar. Rec., 15* (94), 19-24.

Swithinbank, C. W. M. 1973. Higher resolution satellite pictures. *Polar Rec., 16,* (104), 739-741.

Swithinbank, C. W. M. 1973. Satellite view of McMurdo Sound, Antarctica. *Polar Rec., 16,* (105), 851-872.

Swithinbank, C. W. M. 1974. A new map of Alexander Island, Antarctica. *Polar Rec., 17,* (107), 155-157.

Swithinbank, C. W. M. 1974. An international glaciological programme for the Antarctic Peninsula: report of a meeting held in Cambridge, England, 27-30 April 1973. *Polar Rec., 17,* (106), 86-98.

Thomas, R. H. 1970. Survey on moving ice. *Surv. Rev., 20,* (157), 322-338.

Thomas, R. H. 1971. Flow law for Antarctic ice shelves. *Nature, 232,* (30), 85-87.

Thomas, R. H. 1973. The creep of ice shelves: theory. *J. Glaciol., 12,* (64), 45-53.

Thomas, R. H. 1973. The creep of ice shelves: interpretation of observed behaviour. *J. Glaciol.*, *12*, (64), 55-70.

Thomas, R. H. 1973. The dynamics of the Brunt Ice Shelf, Coats Land, Antarctica. *BAS Sci. Rep.*, *79*.

Thomas, R. H. and Coslett, P. H. 1970. Bottom melting of ice shelves and the mass balance of Antarctica. *Nature*, *228*, (5266), 47-49.

Wager, A. C. 1972. Flooding of the ice shelf in George VI Sound. *BAS Bulletin*, *28*, 71-74.

Wager, A. C. (In press). Mapping the depth of a valley glacier by radio echo sounding. *BAS Bulletin*.

Walford, M. E. R. 1972. Glacier movement measured with a radio echo technique. *Nature*, *239*, (5367), 95-96.

FRESHWATER BIOLOGICAL ASSOCIATION
The Ferry House, Far Sawrey, Ambleside, Cumbria

The Freshwater Biological Association, grant aided by the Natural Environment Research Council, is an independent body, founded in 1929, to promote fundamental research in the biology of fresh waters.

The Association's interest in hydrology is limited to the context of the effects of water flow and quality on the ecology of natural waters. Work has been done on the chemistry of rain; there is a continuing interest in the origins, concentrations and fates of dissolved substances present in natural waters. Research in palaeolimnology throws light on ecological conditions in post-glacial catchments.

Armitage, P. D., MacHale, A. M. and Crisp, D. C. 1974. A survey of stream invertebrates in the Cow Green basin (Upper Teesdale) before inundation. *Freshwat. Biol.* *4*, 369-98.

Baker, C. D., Bartlett, P. D., Farr, I. S. and Williams. G. I. 1974. Improved methods for the measurement of dissolved and particulate organic carbon in fresh water and their application to chalk streams. *Freshwat. Biol.*, *4*, 467-81.

Bolas, P. M. and Lund, J. W. G. 1974. Some factors affecting the growth of *Cladophora glomerata* in the Kentish Stour. *Wat. Treat. Exam.*, *23*, 25-51.

Casey, H. 1970. The chemical composition of some southern English chalk streams and its relation to discharge. *Yb. Ass. River Auth.*, 1969, 100-13.

Casey, H. and Newton, P. V. R. 1972. The chemical composition and flow of the South Winterbourne in Dorset. *Freshwat. Biol.*, *2*, 229-34.

Casey, H. and Newton, P. V. R. 1973. The chemical composition and flow of the River Frome and its main tributaries. *Freshwat. Biol.*, *3*, 317-333.

Corlett, J. 1972. Water quality in estuarine barrage reservoirs with special reference to Morecambe Bay. Part 2. Biology. *Wat. Treat. Exam.*, *21*, 119-26.

Crisp, D. T. 1970. Input and output of minerals for a small watercress bed fed by chalk water. *J. appl. Ecol.*, *6*, 117-40.

Crisp, D. T. and Le Cren, E. D. 1970. The temperature of three different small streams in north-west England. *Hydrobiologia*, *35*, 305-23.

Ganf, G. G. and Milburn, T. R. 1971. A conductimetric method for the determination of total inorganic and particulate organic carbon fractions in freshwater. *Arch. Hydrobiol. 69,* 1-13.

George, D. G. 1974. Dispersion patterns in the zooplankton populations of a eutrophic reservoir. *J. Anim. Ecol., 43,* 537-51.

Gilson, H. C. 1971. The impact of water abstraction on river ecology. *J. R. Soc. Arts, 119,* 753-5.

Heaney, S. I. 1974. A pneumatically-operated water sampler for close intervals of depth. *Freshwat. Biol., 4,* 103-06.

Heaney, S. I. and Talling, J. F. 1973. The use of turbidity measurements in lakes. *Proc. Challenger Soc., 4,* 203-04.

Hellawell, J. M., Leatham, H. and Williams, G. I. 1974. The upstream migratory behaviour of salmonids in the River Frome, Dorset. *J. Fish Biol., 6,* 729-44.

Lack, T. J. and Lund, J. W. G. 1974. Observations and experiments on the phytoplankton of Blelham Tarn, English Lake District. I. The experimental tubes. *Freshwat. Biol., 4,* 399-415.

Lund, J. W. G., Jaworski, G. H. M. and Bucka, H. 1971. A technique of bioassay of freshwater, with special reference to algal ecology. *Acta Hydrobiol., Krakow, 13,* 235-49.

Macan, T. T. 1974. Running water. *Mitt. int. Verein. theor. angew. Limnol, 20,* 301-21.

Pennington, W. (Mrs T. G. Tutin). 1973. The recent sediments of Windermere. *Freshwat. Biol., 3,* 363-382.

Pennington, W. (Mrs T. G. Tutin) and Lishman, J. P. 1971. Iodine in lake sediments in northern England and Scotland. *Biol. Rev., 46,* 279-313.

Smith, I. R. 1975. Turbulence in lakes and rivers. *Scient. Publs Freshwat. biol. Ass.,* No. 29, 79.

Sutcliffe, D. W. and Carrick, T. R. 1973. Studies on mountain streams in the English Lake District. 1. pH, calcium and the distribution of invertebrates in the River Duddon. *Freshwat. Biol., 3,* 437-462.

Sutcliffe, D. W. and Carrick, T. R. 1973. Studies on mountain streams in the English Lake District. II. Aspects of water chemistry in the River Duddon. *Freshwat. Biol., 3,* 543-560.

Sutcliffe, D. W. and Carrick, T. R. 1973. Studies on mountain streams in the English Lake District. III. Aspects of water chemistry in Brownrigg Well, Whelpside Ghyll. *Freshwat. Biol., 3,* 561-568.

Talling, J. F. 1971. The underwater light climate as a controlling factor in the production ecology of freshwater phytoplankton. *Mitt. int. Verein. theor. angew. Limnol.,* No. 19, 214-43.

Talling, J. F. 1973. The application of some electrochemical methods to the measurement of photosynthesis and respiration in fresh waters. *Freshwat. Biol. 3,* 335-62.

Westlake, D. F. 1973. Aquatic macrophytes in rivers. A review. *Polskie Archwm Hydrobiol., 20,* 31-40.

INSTITUTE OF GEOLOGICAL SCIENCES
Exhibition Road, London, SW7 2DE

The Institute of Geological Sciences was formed in 1965 by the incorporation of the Geological Survey of

Great Britain with the Overseas Geological Surveys. The scope of its activities ranges widely but within the field of hydrology, the work of its Hydrogeological Department, the successor to an original 'Water Department' set up in 1935, is relevant.

Broadly the Hydrogeological Department is responsible for all aspects of fundamental research (and many applied) into the behaviour of groundwater, the occurrence of aquifers, and the physio-chemical interactions between the two in both unsaturated and saturated zones. The Department's research programme comprises a number of basic long-term projects, mainly within the United Kingdom, which evolve in response to discovery, and an increasing proportion of shorter-term work of a more applied nature, both at home and overseas, commonly commissioned by customer Government Departments in response to particular requirements. Resulting from the implementation of the Rothschild proposals, approximately one third of the programme is funded by DES and the remainder almost equally shared between two major customers, the Department of the Environment and the Ministry of Overseas Development.

The Department has statutory responsibility for the collation of the National Well Record Collection which forms the data base for much of its research work and which is currently being data banked. Processed hydrogeological data is published as Well Inventories and in interpreted form as hydrogeological maps. Fundamental research includes studies of the distribution and characteristics of confined and unconfined aquifers, of the infiltration, zonal behaviour, regional distribution and quality of groundwater, variations in the physical properties of strata and investigations into basin-wide single and multiple aquifers. Applied studies, which may include an appreciable regional component when resource development is involved, also range into local and national pollution and engineering problems. For instance, current applied investigations within the UK include studies into the major problems of groundwater pollution by nitrates and by leachates from landfills; work overseas includes an investigation, in collaboration with Land Resources Division of ODM, into the groundwater resources of two widely-differing areas of the Yemen Republic, numerous short-term resource studies in many developing countries, and the supervision of exploratory drilling for high-enthalpy geothermal resources in the Caribbean.

Anon. (Day, J. B. W., *et al*). 1970. International legend for hydrogeological maps. 101pp. UNESCO Paris. IGS London.

Anon. 1970. Development of shallow non-consolidated aquifers. Report of informal discussion Hydrological Group. *Proc. Instn. civ. Engrs, 47,* 449-455.

Buchan, S. 1968. Estimation of groundwater balances in England. *Bull. int. Ass. scient. Hydrol., 13,* 126-134.

Day, J. B. W., *et al.* 1967. Hydrogeological map of North and East Lincolnshire. *Inst. Geol. Sci., Lond.*

Day, J. B. W., *et al.* 1968. Hydrogeological map of the Dartford (Kent) district. *Inst. Geol. Sci., Lond.*

Day, J. B. W., *et al.* 1970. Hydrogeological map of the Chalk and Lower Greensand of Kent. *Inst. Geol. Sci., Lond.*

Day, J. B. W. 1972. A recent case history of groundwater pollution by organic solvents. Symposium Volume. Groundwater Pollution Working Conference, Water Research Association, Medmenham.

Day, J. B. W., *et al.* (In press). International hydrogeological map of Europe. Sheet B4 (London).

Downing, R. A., Land, D. H., Allender, R., Lovelock, P.E.R. and Bridge, L. R. 1970. The hydrogeology of the Trent River basin. *Hydrogeological Report No. 5. Inst. Geol. Sci., Lond.*

Edmunds, W. M. 1971. Hydrogeochemistry of groundwaters in the Derbyshire Dome with special reference to trace constituents. IGS Report Series 71/7. HMSO Lond.

Edmunds, W. M., Giddings, D. R. and Morgan-Jones, M. 1973. The application of flameless atomic absorption in hydrogeochemical analysis. *Atom. Absorp. Newsl., 12,* 45-9.

Edmunds, W. M., Lovelock, P. E. R. and Gray, D. A. 1973. Interstitial water chemistry and aquifer properties in the Upper and Middle Chalk of Berkshire, England. *J. Hydrol., 19,* 21-31.

Edmunds, W. M., Taylor, B. J. and Downing, R. A. 1969. Mineral and thermal water of the United Kingdom. Monograph on mineral and thermal waters of the world. A—Europe. *Int. geol. Congr.,* Prague.

Foster, S. S. D. 1971. Groundwater conditions in the Thames Flood Plain and their relation to the Thames Barrier. In *Appendix 6, Part I, Thames Flood Prevention, Second Report of Studies.* London: Greater London Council. 24pp.

Foster, S. S. D. (In press). Groundwater Storage – Riverflow Relations in a Chalk Catchment. *J. Hydrol.*

Foster, S. S. D. and Crease, R. I. 1972. Evidence and assessment of the nitrate pollution of Chalk water-supplies in East Yorkshire. Symposium Volume. Groundwater Pollution Working Conference. Water Research Association, Medmenham.

Foster, S. S. D. and Crease, R. I. 1974. Nitrate pollution of Chalk groundwater in East Yorkshire – a hydrogeological appraisal. *J. Instn. wat. Engrs., 28,* 178-194.

Foster, S. S. D. and Crease, R. I. (In press). Hydraulic behaviour of the Chalk aquifer in the Yorkshire Wolds. Technical Note. *Proc. Instn. civ. Engrs.*

Foster, S. S. D. and Price, M. (In press). Water supplies from Ulster valley gravels. *Proc. Instn. civ. Engrs.*

Gray, D. A. 1968. Hydrogeological maps. *Proc. geol. Soc., Lond., 1644,* 288-290.

Gray, D. A. 1969. Groundwater conditions in the Pennant Sandstone. In: *A Selection of Technical Reports submitted to the Aberfan Tribunal.* 3 Vols. Welsh Office, HMSO.

Gray, D. A. 1969. Thames flood prevention: first report of studies. Greater London Council, Appendix 4, 173-197.

Gray, D. A. 1971. Closed circuit T.V. as a research tool in hydrology. In *Visual Media in Geography and Geology.* London: British Universities Film Council Ltd.

Gray, D. A. 1974. The Geology and Mineral Resources of Yorkshire, Chapter 18, *Water Resources and Supply*. York. geol. Soc., Leeds.

Gray, D. A., Allender, R. and Lovelock, P. E. R. 1969. The groundwater hydrology of the Yorkshire Ouse river basin. *Hydrogeol. Rep. 4 Inst. Geol. Sci.*, Lond.

Gray, D. A. and Foster, S. S. D. 1972. Urban influences upon groundwater conditions in Thames flood plain deposits of central London. *Phil. Trans. roy. Soc. Lond. A., 272*, 245-72.

Gray, D. A., Mather, J. D. and Harrison, I. B. (In press). Review of groundwater pollution from waste disposal sites in England and Wales, with provisional guidelines for site selection. *Quart. J. engng. Geol.*

Harvey, B. I. 1973. A computer system for storage and retrieval of hydrogeological data from well records. *IGS Report 73/18*. HMSO Lond.

Kitching, R. and Bridge, L. R. (In press). Lysimeter installations in sandstone at Styrrup, Nottinghamshire, United Kingdom. *J. Hydrol.*

Lovelock, P. E. R. 1970. Laboratory measurement of soil and rock permeability. *Technical Communication 2, Inst. Geol. Sci.*, Lond.

Lovelock, P. E. R. (In press). Aquifer properties of Permo-Triassic Sandstones of the UK. *Bull. geol. Surv. G.B.*

Mather, J. D. 1973. Hydrogeological survey techniques. *Surveyor, 141*, (5225), 26-8.

Mather, J. D., Gray, D. A., Allan, R. A. and Smith, D. B. 1973. Groundwater recharge in the Lower Greensand of the London basin – results of tritium and carbon-14 determinations. *Quart. J. Engng Geol., 6*, 141-52.

Mather, J. D., Gray, D. A. and Jenkins, D. G. 1969. The use of tracers to investigate the relationship between mining subsidence and groundwater occurrence at Aberfan, South Wales. *J. Hydrol., 9*, 136-154.

Mather, J. D.. Gray, D. A. and Houston, J. F. T. 1971. Distribution of the flood plain deposits of the Thames between Chiswick and Beckton. *Res. Rep. 4, Inst. Geol. Sci.*, Lond.

Mather, J. D. and Smith, D. B. 1973. Thermonuclear tritium – its use as a tracer in local hydrogeological investigations. *J. Instn Wat. Engs, 27*, 187-96.

Tate, T. K. 1973. Variations in the design of depth samplers for use in groundwater studies. *Wat. and Wat. Eng., 77*, 223.

Tate, T. K., Robertson, A. S. and Gray, D. A. 1970. The hydrogeological investigation of fissure flow by borehole logging techniques. *Quart. J. Engng Geol., 2*, 192-215.

Tate, T. K., Robertson, A. S. and Gray, D. A. 1971. Borehole logging investigations in the Chalk of the Lambourn and Winterbourne valleys of Berkshire. *Res. Rep. 5, Inst. Geol. Sci.*, Lond.

Tate, T. K. and Robertson, A. S. 1971. Investigations in high salinity groundwater at the Woodfield Pumping Station, Wellington, Shropshire. *Res. Rep. 6, Inst. Geol. Sci.*, Lond.

Williams, B. P. J., Downing, R. A. and Lovelock, P. E. R. 1972. Aquifer properties in the Bunter Sandstone of Nottinghamshire, England. *Proc. Section 11, Int. Geol. Cong.*, Montreal, 169-76.

Well Catalogue Series

The following Records of Wells covering the one-inch geological sheets have been published.

ENGLAND AND WALES

	Publication date
Cromer (131) and Mundesley (132) sheets	1969
Aylsham (147) sheet	1968
North Walsham (148) sheet	1970
Huntingdon (187) and Cambridge (188) sheets	1965
Saffron Walden (205) sheet	1965
Sudbury (206) sheet..	1965
Ipswich (207) and Woodbridge (208) sheets	1966
Hitchin (221) sheet	1970
Great Dunmow (222) and Braintree (223) sheets	1965
Colchester (224) and Felixstowe (225) sheets	1969
Epping (240), Chelmsford (241) and Brightlingsea (242) sheets	1965
Windsor (269) sheet	1966
Dartford (271) sheet	1965
Andover (283) sheet..	1968
Basingstoke (284) sheet	1965
Reigate (286) and Sevenoaks (287) sheets	1968
Haslemere (301) and Horsham (302) sheets	1968
Folkestone (305) and Dover A (306) sheets	1966
Lewes (319), Hastings (320) and Dungeness (321) sheets	1966

SCOTLAND

Kirkmaiden (1), Whithorn (2), Stranraer (3), Wigtown (4), Kirkudbright (5) and Annan (6) sheets	1969
Girvan (7), Carrick (8), Maxwelltown (9), Dumfries (10), Langholm (11), Campbeltown (12), S. Arran (13), Ayr (14), Sanquhar (15), Moffat (16), Jedburgh (17) and Morebattle (18) sheets	1968
North Arran and West Kilbride (21), Kilmarnock (22) and Hamilton (23) sheets ..	1972
Peebles (24), Kelso (25), Berwick-upon-Tweed (26), Haddington (33), Eyemouth (34) and North Berwick (41 – East Lothian Portion) sheets	1966
Edinburgh (32) sheet	1965
Loch Lomond (38), Stirling (39) and Crieff (47) sheets	1965
Kinross (40), North Berwick (41), Perth (48) and Arbroath (49) sheets	1967
Blairgowrie (56), Forfar (57), Banchory (66) and Stonehaven (67) sheets ..	1968

Well Inventory Series

Records of wells with dimensions converted to metric units have been compiled by computer and published for the areas of the following one-inch geological sheets.

Norwich (161) sheet	1973
Bury St Edmunds (189) sheet	1973

INSTITUTE OF HYDROLOGY

Maclean Building, Crowmarsh Gifford, Wallingford, Oxon. OX10 8BB

The Institute of Hydrology is a component body of the Natural Environment Research Council. It is responsible for a broad programme of basic research into all aspects of hydrology and also undertakes projects of an applied nature in Britain and overseas. The Institute's research programme is divided into some 30 projects, most of which are listed in Chapter II, Tables 1-5, under the general subject groupings shown below.

Hydrological systems

The Institute was first established to study the rainfall/runoff relationships operating in natural drainage basins. Complete catchment studies to reveal the influence of geomorphology, of geology and of land use on the response of a catchment to rainfall were set up so that mathematical models of catchment behaviour could be derived to improve existing methods of forecasting and prediction of water resources.

The Institute maintains three catchment experiments: Plynlimon in central Wales with the two basins formed by the forested headwaters of the River Severn and the geologically similar, grass-covered, headwaters of the River Wye; a small clay catchment at Grendon Underwood in north Bucks; and a moorland catchment at Coalburn, Northumberland, newly planted with conifers by the Forestry Commission. In addition, data collected for the Institute from the catchment of the River Cam above Dernford Mill by the Great Ouse River Authority (now the Anglian Water Authority, see page 95) provide extra useful material for modelling purposes.

Mathematical modelling of catchment systems

Consistent batches of data from the experimental catchments are used for mathematical modelling of the runoff pattern. These models are based on the physical laws of water movement and permit derivation of the magnitude of future flows during flood and drought conditions and of their frequency of occurrence.

Water resources studies

The main applied project undertaken by the Institute during the period 1970-75 has been the desk study on the magnitude and frequency of UK floods. The report, published in March 1975, will aid civil engineers in the design of dam spillways, bridges and flood protection works as well as being most relevant in dealing with problems of flood control and land drainage. Other applied studies which have important application to the engineering hydrology problems of the executive departments are as follows: A study of regional low flow characteristics to relate basin characteristics to seasonal low flow regimes which is being sponsored by the Department of the Environment, together with the development of improved techniques for predicting runoff volumes and instantaneous flows for partially or totally urban areas. DOE also agreed to support the continuation of the national archive of flow records, initiated by the original Floods Study, and this has now been taken over by the Water Data Unit (see page 44). The Ministry of Agriculture are funding an investigation into the techniques for converting rainfall to runoff, important for prediction and real-time forecasting. The Department is also sponsoring an extension of the flood frequency analysis carried out as part of the Floods Study.

Advisory studies

Since 1966 the Institute's staff complement carries extra posts under the Overseas Development Ministry's 'home-based' scheme to allow involvement in appropriate overseas aid programmes, especially where the opportunity exists for original research. These projects are fully sponsored by ODM but there are other occasions when Institute members work on water resources projects in the UK and overseas as consultants to outside firms and organisations.

Process studies

Evaporation A combined micrometeorological and biological study to establish the evaporation characteristics of tall vegetation has been in progress for several years in Thetford Forest, Norfolk. Measurements from the meteorological equipment mounted on several towers are taken by a computer-controlled automatic acquisition system which applies calibration factors and performs the necessary calculations for drawing up an energy balance.

Measurements have been made of stomatal resistance – the resistance to water vapour flow through the stomatal openings in individual needles. These results are being used to develop models of the resistance behaviour of different age needles under various conditions and are also being compared with canopy resistance estimates deduced from the meteorological data. Measurements made on the roots have produced valuable information about their distribution and contribution to the resistance to water flow from the soil to the needles.

Interception A proportion of the rain falling on a catchment is intercepted by the leaf canopy of the crop and evaporated back into the atmosphere without playing any further part in the water balance of that catchment. Compared with short vegetation, a forest constitutes a very rough aerodynamic surface and the resultant atmospheric turbulence provides an efficient mechanism for the transport of water vapour away from its surface. Two experimental programmes in which intensive measurements are made of gross rainfall, throughfall and stemflow in conjunction with measurements from automatic weather stations, give detailed estimates of the interception loss.

Surface runoff

Morphometric Variables A map and field survey has started on the gross morphology and deposits of the Wye and Severn catchments at Plynlimon. The end product will be maps of slope angle and superficial deposits from which statistical analysis of mapped variables will reveal the significance of the subdivisions of each catchment in determining the timing, volume and routes of runoff.

Runnoff from Impervious Surfaces Detailed experiments on the rainfall-runoff translation of a variety of impervious surfaces such as roads, roofs and gullies are being made using data from an experimental site at Milton Keynes.

Erosion and Sediment Transport Physically based models of erosion and sediment transport are being developed from data collected from sampling sites in both grassed and forested sites at Plynlimon.

Hydrochemistry Trace element composition of natural waters is being investigated to understand the natural chemical processes that determine the chemistry of runoff in upland catchments.

Subsurface flow
Studies of the distribution of soil moisture in space and time are based on the use of neutron probes. This work has important applications in routine soil moisture storage estimates for modelling purposes and for understanding the nature of soil water abstraction by plants which, together with the influence of soil characteristics on transpiration, is very important in water resource management and irrigation schemes. Tensiometric measurements are used in a zero flux plane technique of determining the deep percolation to the groundwater aquifer and as an additional method of measuring evaporation, independent of meteorological energy balance techniques.

Field studies involving the detection and distribution of sub-surface flow in catchment areas are concerned with the objective separation of the runoff hydrograph. Percolation and infiltration processes are investigated by studying the distribution of atmospheric tritium in soils as a means of determining zones of groundwater re-charge in areas of heterogenous soil types of unknown permeabilities.

Physical Models Models of catchment response are being derived from quantitative process studies of representative first order basins and small plots as an alternative to optimised parameter methods. Such physically based models are required for hydrograph prediction for catchments for which there are insufficient records of rainfall and runoff to permit the application of traditional modelling methods. The experimental work is taking place in the Wye and Severn catchments at Plynlimon which has the advantage that model predictions may be checked against measurements obtained by the catchment project described earlier.

Instrumentation
The continued emphasis on mathematical modelling of catchment behaviour means a constant pressure to improve data collection. Over the years, considerable effort has gone into refining existing equipment or to sponsoring the design and manufacture of improved devices. The Wallingford soil moisture neutron probe is known all over the world and the Institute's automatic weather station is proving to be a most valuable reference tool.

Baty, A. J., Rodda, J. C. and Templeman, R. F. 1973. A measurement of albedo changes resulting from afforestation practices. *Hydrol. Sci. Bull., 18* (*4*), 451-459.

Bell, J. P. 1973. Neutron probe practice. *Inst. Hydrol., Wallingford, Rep. 19.*

Beran, M. A. 1973. Estimation of design floods and the problem of equating the probability of rainfall and runoff. *Proc. International symposium on the design of water resources projects with inadequate data, UNESCO/WMO/IAHS,* 41-50.

Beran, M. A. and Sutcliffe, J. V. 1972. An index of flood-producing rainfall based on rainfall and soil moisture deficit. *J. Hydrol., 17,* 229-236.

Blyth, K., Cooper, M. A. R., Lindsey, N. E. and Painter, R. B. 1974. Snow depth measurement with terrestrial photos. *Photogramm. Engng., XL* (8), 937-942.

Blyth, K. and Painter, R. B. 1974. Analysis of snow distribution using terrestrial photogrammetry. *Advanced concepts and techniques in the study of snow and ice resources. Nat. Acad. Sci., Washington* 679-687.

Blyth, K. and Rodda, J. 1973. A stream length study. *Wat. Resour. Res., 9* (5), 1454-1462.

Clarke, R. T. 1971. Objectives and methods of data processing and analysis in the water treatment context. *Wat. Treatm. and Examn, 20,* 95-107.

Clarke, R. T. 1971. The use of the term 'linearity' as applied to hydrological models. *J. Hydrol., 13,* 47-51.

Clarke, R. T. 1973. A review of some mathematical models used in hydrology, with observations on their calibration and use. *J. Hydrol., 19* (*1*), 1-20.

Clarke, R. T. 1974. The representation of a short period of experimental catchment data by a linear stochastic difference equation. *Symp. on mathematical models in hydrology, IASH/UNESCO, Warsaw, 1971*, 3-15.

Clarke, R. T. and Edwards, K. A. 1972. The application of the analysis of variance to mean areal rainfall estimation. *J. Hydrol., 15,* 97-112.

Clarke, R. T. and McCulloch, J. S. G. 1975. Recent work in the comparison of the effects of alternative uses (coniferous forest, upland pasture) on water catchment behaviour. *Proc., Water Space Amenity Conf., Conservation and Land Drainage,* May 1975.

Cole, J. A., McKerchar, A. I. and Moore, R. J. (In press). An on-line forecasting system incorporating radar measurement of rainfall as used to assist the short-term regulation of the River Dee in North Wales. *Symp. on application of mathematical models in hydrology and water resources, Bratislava, 1975. IAHS publication 115.*

Cunnane, C. 1973. A particular comparison of annual maxima and partial duration series methods of flood frequency prediction. *J. Hydrol., 18* (*3/4*), 257-272.

Cunnane, C. and Nash, J. E. 1974. Bayesian estimation of frequency of hydrological events. *Symp. on mathematical models in hydrology, IASH/UNESCO, Warsaw, 1971*, 47-55.

Dickinson, W. T. and Douglas, J. R. 1972. A conceptual runoff model for the Cam catchment. *Inst. Hydrol., Wallingford, Rep. 17.*

Douglas, J. R. 1972. A system for processing data from autographic recorders. *Inst. Hydrol., Wallingford, Rep. 14.*

Douglas, J. R. 1974. Conceptual modelling in hydrology. *Inst. Hydrol., Wallingford, Rep. 24.*

Edwards, K. A. 1972. A note on the 'Calculation of mean areal depth of precipitation'. *J. Hydrol, 15,* 171-173.

Edwards, K. A. and Rodda, J. C. 1970. A preliminary study of the water balance of a small clay catchment. *Symp. on the results of research on representative and experimental basins, IASH/UNESCO,* Wellington, *J. Hydrol., New Zealand, 9,* 202-218.

Farquharson, F. A. K., Lowing, M. J. and Sutcliffe, J. V. (In press). Some aspects of design flood estimation. *Symp. on inspection, operation and improvement of existing dams. BNCOLD/University of Newcastle symposium,* September 1975.

Frost, J. and Clarke, R. T. 1972. Estimating the T-year flood by the extension of records of partial duration series. *Bull. int. Ass. hydrol. Sci., 17,* 209-217.

Frost, J. and Clarke, R. T. 1973. Use of cross correlation between hydrological time series to improve estimates of lag one autoregressive parameters. *Wat. Resour. Res., 9* (*4*), 906-917.

Gash, J. H. C. and Stewart, J. B. (In press). The average surface resistance of a pine forest derived from Bowen ratio measurements. *Bound. Lay. Meteorol.*

Gay, L. W. and Stewart, J. B. 1974. Energy balance studies in coniferous forests. *Inst. Hydrol., Wallingford, Rep. 23.*

Herbertson, P. W., Douglas, J. R. and Hill, A. 1971. River level sampling periods. *Inst. Hydrol., Wallingford, Rep. 9.*

Jansen, J. M. L. and Painter, R. B. 1974. Predicting sediment yield from climate and topography. *J. Hydrol., 21* (*4*), 371-380.

Leese, M. N. 1973. The use of censored data in estimating T-year floods. *Proc. International symposium on the design of water resources projects with inadequate data, UNESCO/WMO/IAHS,* 236-247.

Leese, M. N. 1973. Use of censored data in the estimation of Gumbel distribution parameters for annual maximum flood series. *Wat. Resour. Res., 9* (6), 1534-1543.

Leese, M. N., Newson, A. J. and Clarke, R. T. 1975. Analysis of data from Plynlimon raingauge networks: April 1971 – March 1973. *Inst. Hydrol., Wallingford, Rep. 27.*

Leese, M. N., Strangeways, I. C. and Templeman, R. F. 1973. Field performance of Institute of Hydrology automatic weather station. *Inst. Hydrol., Wallingford, Rep. 21.*

Lowing, M. J. and Newson, M. D. 1973. Flood event data collation. *Wat. and Wat. Engng, 77*, 91-95.

McCulloch, J. S. G. and Painter, R. B. 1973. The application of multi-spectral scanning to hydrology. *ESRO, Multispectral scanning systems and their potential application to earth resource surveys, Vol. 4: Earth Science applications,* 127-151.

Mandeville, A. N. and O'Donnell, T. 1973. The introduction of time variance to linear conceptual catchment models. *Wat. Resour. Res., 9* (*2*), 298-310.

Mandeville, A. N. and Rodda, J. C. 1970. A contribution to the objective assessment of areal rainfall amounts. *Symp. on the results of research on representative and experimental basins, IASH/UNESCO Wellington, J. Hydrol., New Zealand, 9,* 281-291.

Natural Environment Research Council 1975. Flood Studies Report. 5 vols.

Newson, M. D. 1973. The carboniferous limestone of the UK as an aquifer rock. *Geogr. J., 139* (*2*), 294-310.

Newson, M. D. 1975. Mapwork for flood studies. Part 1: Selection and derivation of indices. *Inst. Hydrol., Wallingford, Rep. 25.*

Newson, M. D. 1975. Plynlimon floods of August 5th/6th 1973. *Inst. Hydrol, Wallingford, Rep. 26.*

Newton, S. G. and Painter, R. B. 1974. A mathematical examination of urban runoff prediction. *Proc. Instn. civ. Engrs. Part 2, 57,* 143-157.

Oliver, H. R. 1971. Wind profiles in and above a forest canopy. *Quart. J. R. Met. Soc., 97,* 548-553.

Oliver, H. R. 1975. Ventilation in a forest. *Agric. Meteorol., 14,* 347-355.

Oliver, S. A. and Oliver, H. R. 1973. A computer method for automatic acquisition of wind speed data. *J. Phys., E., 6,* 401-403.

Painter, R. B. 1971. The hydrograph time-to-peak method of flood prediction. *Wat. & Wat. Engng, 75,* 235-237.

Painter, R. B. 1972. The measurement of bed load movement in rivers. *Wat. and Wat. Engng, 76,* 291-294.

Painter, R. B. 1972. Some present uses of remote sensing in monitoring hydrological variables. Presented at *Symp. on remote sensing, University of Bristol. Environmental remote sensing, E. C. Barrett and L. F. Curtis, eds. Edward Arnold, 1974.*

Painter, R. B. 1973. The application of multispectral scanning to the measurement of water depth and quality. *ESRO Multispectral scanning systems and their potential application to earth resource surveys, Vol. 4: Earth science applications,* 136-139.

Painter, R. B., Blyth, K., Mosedale, J. C. and Kelly, M. 1974. The effect of afforestation on erosion processes and sediment yield. *Symp. on effects of man on the interface of the hydrological cycle with the physical environment. IAHS/UNESCO/WMO,* Paris *1974,* 62-68.

Plinston, D. T. 1972. Parameter sensitivity and interdependence in hydrological models. In : *Mathematical models in ecology, the 12th Symposium of the Brit. Ecol. Soc. 1971* (ed. J. N. R. Jeffers). Blackwell Scientific Publications, Oxford, 237-247.

Plinston, D. T. and Hill, A. 1974. A system for the quality control and processing of streamflow, rainfall and evaporation data. *Inst. Hydrol., Wallingford, Rep. 15.*

Richards, D. 1971. A computer program to use orthogonal polynomials in two variables for surface-fitting. *Inst. Hydrol., Wallingford, Rep. 13.*

Richards, D. 1975. The effects of reducing rain-gauge network density on goodness of conceptual model fit and prediction. *Inst. Hydrol., Wallingford Rep. 28.*

Roberts, G. 1972. The processing of soil moisture data. *Inst. Hydrol., Wallingford, Rep. 18.*

Rodda, J. C. 1970. Controlling flood and drought. *Geogr. Mag., 43,* 112-115.

Rodda, J. C. 1970. On the questions of rainfall measurement and representativeness. *Symp on World Water Balance, IASH/UNESCO, Reading,* 173-186.

Rodda, J. C. 1970. Rainfall excesses in the United Kingdom. *Trans. Inst. Brit. Geogr., 49,* 49-60.

Rodda, J. C. 1971. Drought, agriculture and hydrology, Ch. 7 in *Assessment of Drought* (Ed. C. E. Hounam), Report of the Commission for Agricultural Meteorology Working Group to CAgM V, Document 22. Geneva.

Rodda, J. C. 1971. The precipitation measurement paradox – the instrument accuracy problem. *WMO/IHD Project Rep. 16 Wld met Orgn, Geneva*, 42p.

Rodda, J. C. 1971. Report on precipitation. *Bull. int. Ass. scient. Hydrol., 16*, (*4*), 37-47.

Rodda, J. C. 1972. Precipitation. Ch. 1 – 1.2 in: *WMO casebook on hydrological network design practice*, WMO No. 324.

Shuttleworth, W. J. 1975. The concept of intrinsic surface resistance: energy budgets at a partially wet surface. *Bound. Lay. Meteorol., 8* (*1*), 81-99.

Stewart, J. B. 1971. The albedo of a pine forest. *Quart. J. R. met. Soc., 97*, 561-564.

Stewart, J. B. and Oliver, S. A. 1972. Evaporation studies at Thetford, Norfolk. In: *Research papers in forest meteorology, an Aberystwyth Symposium, Ed J. A. Taylor* (Based on Symposium 13, March, 1970). Cambrian News, Aberystwyth Ltd., 91-99.

Stewart, J. B. and Thom, A. S. 1973. Energy budgets in a pine forest. *Quart. J. R. met. Soc., 99* (*419*). 154-170.

Strangeways, I. C. 1972. Automatic weather stations for network operation. *Weather, 27*, 403-408.

Strangeways, I. C. and Lisoni, L. 1973. Long-distance telemetry of data for flood forecasting. *UNESCO, Nature and Resources, IX* (*4*), 18-21.

Strangeways, I. C. and Templeman, R. F. 1974. Logging river level on magnetic tape. *Wat. Services, 78* (*936*), 57-60.

Thom, A. S., Stewart, J. B., Oliver, H. R. and Gash, J. H. C. 1975. Comparison of aerodynamic and energy budget estimates of fluxes over a pine forest. *Quart. J. R. Met. Soc., 101*, 93-105.

Truesdale, V. W. 1973. The factors controlling the natural chemical composition of rivers – a critical review. *Proc. Challenge. Soc., IV* (*5*).

Truesdale, V. W. and Smith, P. J. 1975 The automatic determination of iodide or iodate in solution by catalytic spectrophotometry, with particular reference to river water. *Analyst, 100*, 111-123.

SCIENCE RESEARCH COUNCIL

APPLETON LABORATORY

Ditton Park, Slough SL3 9JX

The station has an interest in the characteristics of precipitation since they affect the absorption and scattering of radio and optical waves. Studies of the instantaneous local variations in rainfall over short distances and of drop-size distributions are of special importance. Such studies are in progress and are being combined with simultaneous measurements of radar back-scatter and of millimetre, sub-millimetre and infra-red transmissions. A rapid-response raingauge has been developed for this work, particularly for use in a network of gauges in a computer-controlled system linked by telemetry.

Radio and Space Research 1971/73 – published by SRC.

MINISTRY OF AGRICULTURE, FISHERIES AND FOOD

LAND DRAINAGE, WATER SUPPLY AND MACHINERY DIVISION

Great Westminster House
Horseferry Road, London SW1P 2AE

Field Drainage Experimental Unit
Anstey Hall, Maris Lane, Trumpington
Cambridge CB2 2LF

The Ministry of Agriculture, Fisheries and Food gives grant aid to Regional Water Authorities, Drainage Boards and Local Authorities towards the cost of flood protection and land drainage schemes for rivers and other water courses. The Ministry also grant aid schemes for field drainage of land and farm water supplies including the irrigation of agricultural crops.

The professional and technical matters involved in the work of this administrative division are dealt with by the Drainage arm of the Agricultural Development and Advisory Service. This includes the Field Drainage Experimental Unit at Cambridge which is concerned with problems in field drainage and in particular with design standards. The hydrological interest of the Unit is in the movement of soil water, the behaviour of drainage installations and the development of mathematical models for predictive use. Most effort is on field scale hydrology but some work is done on larger catchments. Hydrological data are collected from a number of experimental sites and extensive soil water regime studies are being made on sites throughout England and Wales. The Unit does not undertake fundamental research but is in close contact with research centres in Britain and overseas.

FDEU Annual Report 1969.
" " " 1970.
" " " 1971.
" " " 1972.
" " " 1973.

Berryman, C., *et al.* 1974. Hydraulic conductivity and infiltration rates. *FDEU Tech. Bull. 74/4.*

Carter, C. and Trafford, B. D. 1973. Idealized soil profiles and corresponding drainage designs. *FDEU Tech. Bull., 73/12.*

Carter, C., Hawes, M. and Trafford, B. D. 1974. An assessment of the damage caused by laying new drains through a growing crop. *FDEU Tech. Bull., 74/14.*

Croote, S. 1974. The use of plastics in field drainage. *FDEU Tech. Bull., 74/10.*

Dennis, C. W. 1973. The hydraulic characteristics of plastics land drainage pipe. Tech. Note 75. *Proc. Inst. of Civil Engineers,* March 1973.

Dennis, C. W., Massey, W., Rands, J. G., Rycroft, D. W. and Trafford, B. D. 1974. Permeable backfill over drains. Papers from the Stoneleigh Conf., October 1973. *FDEU Tech. Bull., 74/11.*

Massey, W. 1973. Drainage economics – change in land use as a result of drainage. *FDEU Tech. Bull., 73/8.*

Parker, A. 1972. The probability of rainfalls significant to field drainage. *FDEU Tech. Bull., 72/9.*

Pullen, D. W. 1973. An analysis of the present use of permeable backfill. *FDEU Tech. Bull., 73/1.*

Rands, J. G. 1973. Drainage design and the soils of Eastern Region. *FDEU Tech. Bull. 73/3.*

Rands, J. G. 1974. An analysis of under-drainage costs in England and Wales. *Farm Contractor,* Sept. 1974.

Rycroft, D. W. 1972. A literature review on the background to mole drainage. *FDEU Tech. Bull., 72/10.*

Rycroft, D. W. 1973. Deep seepage on a boulder clay soil. Results of an investigation at Boxworth. *FDEU Tech. Bull., 73/7.*

Rycroft, D. W. and Thorburn, A. A. 1974. Water stability tests on clay soils in relation to mole draining. *Soil Sci., 117,* (*5*), 306-310.

Rycroft, D. W. and Alcock, M. 1974. The assessment of suitability of soils for moling. *FDEU Tech. Bull., 74/1.*

Smitham, G. M. 1972. A survey of the conditions under which permeable backfill is used. *FDEU Tech. Bull., 72/8.*

Steinhardt, R. and Trafford, B. D. 1974. Some effects of sub-surface drainage and ploughing on the structure and compactability of a clay soil. *J. Soil Sci., 25* (*2*), 138-152.

Trafford, B. D. 1972. Agricultural land drainage design. *Proc. Welsh Soils Discussion Group.*

Trafford, B. D. 1972. The application of synthetic materials to field drainage techniques and practices. *Paper R3, Q28.1 8th Cong. ICID Varna.*

Trafford, B. D. 1972. The measurement of water tables. *FDEU Tech. Bull., 72/3.*

Trafford, B. D. 1972. The drainage experiment on radar restored opencast coal site. *FDEU Tech. Bull., 72/6.*

Trafford, B. D. 1972. The theoretical background to the use of gravel as the material on and around field drains. *FDEU Tech. Bull., 72/7.*

Trafford, B. D. and Rycroft, D. W. 1973. Observations on the soil water regimes in a drained clay soil. *J. Soil Sci., 24,* (*3k*), 380-391.

Trafford, B. D. 1973. Farm drainage. *J. Royal Soc. of Arts,* Feb. 1973, 134-145.

Trafford, B. D. 1973. The analysis of inflow into some experimental agricultural drains. *Hydrol. Sci. Bull., 18,* (*4*), 431-440.

Trafford, B. D. 1973. The relationship between field drainage and arterial drainage – theoretical aspects. *FDEU Tech. Bull., 73/10.*

Trafford, B. D. and Rands, J. G. 1974. The relationship between drainage design rate and rainfall. *FDEU Tech. Bull., 74/5.*

Trafford, B. D. 1974. A guide to drainage design technique based on scientific methods. *FDEU Tech. Bull., 74/7.*

Trafford, B. D. and Rycroft, D. W. 1974. The natural soil as trench backfill. *FDEU Tech. Bull., 74/9.*

Trafford, B. D. and Dennis, C. W. 1974. The role of the pipe in field drainage. *FDEU Tech. Bull. 74/15.*

AGRICULTURAL DEVELOPMENT AND ADVISORY SERVICE

Great Westminster House, Horseferry Road, London SW1P 2AE

Soil Scientists of the Agricultural Development and Advisory Service at several centres in England and Wales carry out work on soil moisture. This includes the determination of moisture characteristics and the measurement of the heights of water tables on a range of soils. Recently, with increasing intensification of farming, there has been considerable interest in moisture status in relation to the structural damage which can be caused in soils by cultivation or grazing under wet soil conditions.

MINISTRY OF DEFENCE

HYDROGRAPHIC DEPARTMENT

Creechbarrow House, Taunton, Somerset

The Hydrographic Department carries out hydrographic surveys around the British Isles and British territories abroad. It produces charts to cover the whole world, for issue to the Royal Navy and for sale to British, Commonwealth and foreign merchant fleets; it also publishes the necessary sailing directions, tide tables, radio and light lists to complement this world coverage; surveys may also be carried out in the open oceans and, as mutually arranged, in the waters around certain Commonwealth and foreign territories.

The Tidal Branch of the Hydrographic Department is responsible for the publication of the Admiralty Tide Tables which give world wide tidal predictions. For this reason it is interested in the effect of river water and seasonal variations on tidal heights.

The Storm Tide Warning Service, a branch of the Hydrographic Department, is responsible for issuing warnings to the appropriate authorities of abnormally high sea levels on the east coast of England. In order that it may have the best possible service on the meteorological situation the office is situated within the Meteorological Office Headquarters, at Bracknell. Research into the improvements of the accuracy of the warnings of the service is continuously carried out by its staff officers.

In addition, an experimental 'Negative Surge Warning Service' has been started to forecast occasions when meteorological conditions produce tides far enough below the predicted values to constitute a navigational hazard. Initial empirical equations for this purpose were derived by the staff of the Storm Tide Warning Service. Research into methods of forecasting a more complex form of this surge have now been undertaken by Southampton University Department of Oceanography, directed by the STWS.

METEOROLOGICAL OFFICE

London Road, Bracknell, Berks, RG12 2SZ

The Meteorological Office is responsible for providing a meteorological service to other government departments, aviation, shipping, agriculture, commerce, industry and the general public and for research in meteorology and in other related fields of geophysics. Its function also includes the collection, distribution and publication of meteorological information, mostly from the United Kingdom. In particular the hydrometeorological section of the Meteorological Office is concerned with the measurement, collection, quality control, dissemination and analysis of precipitation data and for estimates of evaporative water losses and soil moisture deficits.

Two major cooperative research projects deserve special mention, namely the Dee Weather Radar Project and the Flood Studies Project. The Dee Weather Radar Project is concerned with the development of a radar system, including a computer, to measure rainfall in real-time on a space and time-scale appropriate to hydrological applications in the upper catchments of the River Dee, North Wales. A comparison of areal rainfalls derived from radar measurements with areal rainfalls determined from a dense network of recording raingauges has shown that the radar is sufficiently accurate for these applications; additionally it has important observational advantages in thunderstorm situations and has potential short-period forecasting value.

The Meteorological Flood Study Project was completed in 1974 with the production of Volume 2 of the Flood Studies Report, published by NERC in 1975 (see page 59). The results have a direct application to storm drainage design, to river management and to the control and alleviation of flooding. The meteorological volume deals with the determination of the depth-duration-return period relationships for point rainfalls of any duration at any location in the United Kingdom. Parallel relationships are available for areal rainfalls. Other aspects which are dealt with include maximum rainfalls of any duration, duration-intensity relationships of rainfall and the contribution of snowmelt to river flooding. Other projects include the development of a much improved operational system for the quality control of rainfall data using a computer. One result of this improved system is that it now takes only about 30 minutes on the powerful IBM 360/195 computer to quality control daily rainfall data for nearly 7000 United Kingdom stations each month. Studies of rainfall deficiency in the United Kingdom are in progress to provide a method of determining the return period and the spatial variability of dry weather. Following a number of investigations, work is in progress on improving a computer model for estimating evaporative water loss and soil moisture deficit on an operational basis.

Allen, P. G. 1972. The routine processing of current rainfall data by computer. *Met. Mag.*, Lond., *101*, 344-345.

Bleasdale, A. and Chan, Y. K. 1972. Orographic influences on the distribution of precipitation. *Proc. Symp. Distribution of precipitation in mountainous areas.* Geilo. WMO 326, Vol. 2, 322-333.

Bleasdale, A. 1973. Meteorological contributions to operational hydrological forecasting in the United Kingdom. *Met. Mag.*, Lond., *102*, 298-308.

Clarkson, C. S. 1971. On the performance of various types of rain-gauges in the field. *Met. Mag.*, Lond., *100*, 241-255.

Collier, C. G., Harrold, T. W. and Nicholass, C. A. 1975. A comparison of areal rainfall as measured by a raingauge-calibrated radar system and rainguage networks of various densities. Proc. 16th Radar Met. Conf., Houston, Texas.

Crawford, S. G. 1972. A recording gravimetric rain-gauge – towards an absolute reference instrument. *Met. Mag.*, Lond., *101*, 368-374.

Gloyne, R. W. 1971. A note on the measurement and estimation of evaporation. *Met. Mag.*, Lond., *100*, 322-328.

Grindley, J. 1970. Estimating and mapping evaporation. *Proc. Symp. World Water Balance*. Reading, UNESCO (IASH). 92, 200-213.

Harrold, T. W. and Nicholass, C. A. 1972. The accuracy of some recent radar estimates of surface precipitation. *Met. Mag.*, Lond., *101*, 193-205.

Harrold, T. W., Bussell, R. and Grinstead, W. A. 1972. The Dee Weather Radar project. *Proc. Symp. Distribution of precipitation in mountainous areas*. Geilo. W.M.O. 326. Vol. 2, 47-61.

Harrold, T. W., English, E. J. and Nicholass, C. A. 1974. The accuracy of radar-derived rainfall measurements in hilly terrain. *Quart. J. R. Met. Soc.*, *100*, 331-350.

Holgate, H. T. D. 1973. Rainfall forecasting for river authorities. *Met. Mag.*, Lond., *102*, 33-48.

Holland, D. J. and Noad, J. M. 1970. Mapping spatially smoothed rainfall. *Met. Mag.*, Lond., *99*, 38-45.

Hopkins, J. S. 1972. A study of snowmelt floods in a mountainous catchment using limited meteorological data. *Met. Mag.*, Lond., *101*, 221-228.

Jackson, M. C. 1974. Largest two-hour falls of rains in the British Isles. *Weather*, Lond., *29*, 71-73.

Jackson, M. C. 1975. Annual duration of any rainfall intensity. *Met. Mag.*, Lond., *104*, 243-248.

Lawrence, E. N. 1971. Synoptic type rainfall averages over England and Wales. *Met. Mag.*, Lond., *100*, 333-339.

Meteorological Office, 1975. U.K. Flood Studies Report, Vol. 2. Meteorological Studies, NERC, Lond.

Murray, R. 1971. Forecasting March temperature and rainfall for England and Wales. *Met. Mag.*, Lond., *100*, 161-173.

Murray, R. 1972. An objective method of foreshadowing winter rainfall and temperature for England and Wales. *Met. Mag.*, Lond., *101*, 97-110.

Murray, R. 1972. Prediction of summer rainfall and temperature for England and Wales from anomalous atmospheric circulation in spring. *Met. Mag.*, Lond., *101*, 253-264.

Murray, R. 1973. Forecasting seasonal rainfall and temperature for England and Wales in spring and autumn from anomalous atmospheric circulation over the northern hemisphere. *Met. Mag.*, Lond., *102*, 15-26.

Murray, R. 1974. Indicators of monthly mean temperature and rainfall for England and Wales based on antecedent monthly pressure anomalies over the northern hemisphere. *Met. Mag.*, Lond., *103*, 70-73.

Painter, H. E. 1975. Preliminary results from a gravimetric rain-gauge. *Met. Mag.*, Lond., *104*, 69-78.

Shearman, R. J. 1975. Computer quality control of daily and monthly rainfall data. *Met. Mag.*, Lond., *104*, 102-108.

Stewart, J. B. and Thom, A. S. 1973. Energy budgets in a pine forest. *Quart. J. R. Met. Soc., 99*, 154-170.

Wales-Smith, B. G. 1971. Monthly and annual totals of rainfall representative of Kew, Surrey, from 1697 to 1970. *Met. Mag.*, Lond., *102*, 157-171.

Wales-Smith, B. G. 1973. Evaporation in the London area from 1698 to 1970. *Met. Mag.*, Lond., *102*, 281-291.

Wales-Smith, B. G. 1974. Estimates and measurements of potential evaporation and evapotranspiration for operational hydrology. *British Rainfall 1967*. HMSO Lond. 203-211.

DEPARTMENT OF THE ENVIRONMENT

BUILDING RESEARCH ESTABLISHMENT

Garston, Watford, WD2 7JR, Hertfordshire

The Building Research Station is concerned with the study of the problems of building and allied branches of civil engineering.

The Geotechnics Division deals with problems of the behaviour of the ground in relation to civil engineering construction. It studies the influence of groundwater and soil moisture on the stability and deformation of the ground, natural slopes, excavations, earthworks, foundations and dams. Some work is in progress on the seasonal and long term movements of the peat in Holme Fen.

DIRECTORATE GENERAL WATER ENGINEERING

2 Marsham Street, London, SW1P 3EB

The Directorate General Water Engineering was set up in 1970 to advise the Secretary of State among other things on aspects of water supply, waste water treatment and waste disposal; to sponsor research and development projects in these fields and to provide advice and assistance to the water and waste industries.

RESEARCH AND DEVELOPMENT DIVISION

Within these broad terms of reference the Directorate through its Research and Development Division sponsors a large programme of projects with local authorities and research stations. The research objectives are to further knowledge and to identify areas where further research should be concentrated. Development work aims to provide a cheaper, safer and more effective water supply and disposal industry.

The Research and Development Division is involved with hydrological research into a number of projects concerned with the design and performance of sewerage systems. Studies of the effects of separating foul sewage and surface water drainage have been initiated together with maintaining the quality and frequency of storm overflow discharges in combined systems.

A four year study of sewer flows in various sizes and types of catchment has also been made with the object of establishing the relationship between rainfall and sewer flows. It is expected that this project will provide useful data for the present study by the Institute of Hydrology into urban run-off and for the DOE Working Party on the Hydrological Aspects of Storm Sewer

Design. Incidental aims of these projects are to develop and improve the collection of field data and to evaluate existing instrumentation and data analysis.

The Division has also collaborated in the study of the effects of urbanisation on run-off in a river catchment in Gloucestershire and is promoting a survey of groundwater from waste tips using natural tritium as an indicator.

HYDRAULICS RESEARCH STATION
Howbery Park, Wallingford, Oxon, OX10 8BA

The Hydraulics Research Station, Wallingford, is a research establishment of the Department of the Environment and carries out hydraulics research of importance to civil engineers and hydrologists. This research is primarily concerned with the free surface flow of water in open channels and, to a lesser extent in closed conduits. It extends from the small stream to the sea within the hydrological cycle.

The research is directed mainly to be of assistance in civil engineering design and construction. It comprises basic (background) research to extend scientific knowledge of free surface hydraulics and *ad hoc* investigations of important practical problems. Hydrological considerations enter into much of the work.

The problems investigated, both basic and *ad hoc*, are concerned with the hydraulics of streams, rivers, estuaries, ports, harbours, coastal defence and the hydraulic structures associated with them. Research of special hydrological interest within this range includes that on the hydraulic characteristics of measuring weirs and flumes, stage discharge relationships of fixed and loose boundary open channels of many kinds, flood alleviation and control, flood discharge structures, dams and reservoirs, flood wave propagation, tidal flow alone and in combination with freshwater flow, density current phenomena and sediment transport and associated problems over the whole range of the Station's work.

There is also a unit at the Station, funded by the Ministry of Overseas Development, in which basic research is carried out into problems arising in developing countries in the fields of water resources and irrigation.

Ackers, P. 1970. Meandering of small streams in alluvium. *HRS Rep. INT 77.*

Ackers, P. 1971. Similarity in sediment transport. *HRS Report INT 90.*

Ackers, P. 1971. Sediment transport in channels: an alternative approach. *HRS Report INT 102.*

Ackers, P. 1971. Flow measurement by weirs and flumes. *Paper presented at International conference on modern developments in flow measurement, AERE Harwell, 21st-23rd Sept. 1971.*

Ackers, P. and Charlton F. G. 1970. Dimensional analysis of alluvial channels with special reference to meander length. *J. of Hydraul. Res. 8* (*3*), 287-316.

Ackers, P. and Charlton, F. G. 1970. Meander geometry arising from varying flows. *J. Hydrol., XI* (*3*), 230-252.

Ackers, P. and Charlton, F. G. 1970. The geometry of small meandering streams. *Proc. I.C.E. Supplement* (*xii*) (*Paper 7328s*).

Ackers, P. and Charlton, F. G. 1970. The slope and resistance of small meandering channels. *Proc. I.C.E. Supp.* (*Paper 7362s*).

Ackers, P. and White, W. R. 1973. Sediment transport. A new approach and analysis. *Proc. ASCE, 99 HY11*, 2041-2060.

Chambers, D. N. and Rogers, K. G. 1973. The economics of flood alleviation. *L.G.O.R.U. Report No. C155.*

Charlton, F. G. 1974. The effect of meandering on the behaviour of alluvial channels. *Paper presented at Institute of British Geographers Applied Environmental Sciences Seminar, University of East Anglia, Norwich.*

Crabbe, A. D. 1974. Some hydraulic features of the square-edged broad crested weir. *Wat. Serv. 78* (*944*), 354-8.

Grindley, J. 1970. Calibration and behaviour of current meters. *HRS Rep. INT 80.*

Grindley, J. 1970. Calibration and behaviour of current meters: effect of oblique flow. *HRS Rep. INT 87.*

Grindley, J. 1971. Behaviour and calibration of current meters: effect of suspension. *HRS Rep. INT 93.*

Grindley, J. 1971. Behaviour and calibration of current meters: accuracy. *HRS Rep. INT 95.*

Grindley, J. *1971*. Calibration and behaviour of current meters: drag. *HRS Rep. INT 96.*

Grindley, J. 1972. Calibration and behaviour of current meters: tests in flowing water. *HRS Rep. INT. 99.*

Harrison, A. J. M. 1971. The influence of valley gravels on the movement of a flood wave. *Proc. 14th Congress I.A.H.R., Paris, 5, 1971*, 187-191.

Harrison, A. J. M. 1975. Open channel hydraulics. *Conf. on engineering hydrology today, Inst. civ. Engrs., Feb. 1975.* 45-51.

Kawecki, W. 1973. Boundary roughness for natural channels. *HRS Rep. INT 120.*

Kiff, P. R. 1973. Particle size analysis of sediments. *Lab. Prac., 22* (*4*).

Kiff, P. R. 1973. Problems associated with the determination of a size distribution by using sieves with particular reference to the sizing of marine sediments. *Proc. Soc. for analyt. chem., 10* (*5*), 114-115.

Lean, G. H. 1971. The settling velocity of particles in channel flow. *HRS Rep. INT 91.*

Owen, M. W. 1970. Properties of a consolidating mud. *HRS Rep. INT 83.*

Owen, M. W. 1970. A detailed study of the settling velocity of an estuary mud. *HRS Rep. INT 78.*

Owen, M. W. 1971. The effect of turbulence in the settling velocity of silt floes. *Proc. XIV Congress I.A.H.R.*, 27-32.

Owen, M. W. 1972. The effect of temperature on the settling velocities of an estuary mud. *HRS Rep. INT 106.*

Perkins, J. A. and Wooldridge, G. 1974. Calibration programs for critical depth flumes. *HRS Rep. INT 132.*

Price, R. K. 1973. Numerical models for some problems in water systems. *Paper presented to Water Research Association Conference on 'Computer uses in water systems' Sept. 1973*, 3.

Price, R. K. 1970. Flood routing in natural rivers. *Proc. I.A.H.R. 15th Congress, Istanbul, Turkey 1973*(*4*), 139-144.

Price, R. K. 1973. Flood routing methods for British rivers. *Proc. ICE, 55* (*2*), 913-930.

Price, R. K. 1974. Comparison of four numerical methods for flood routing. *Proc. ASCE HY7*, 879-99.

Price, R. K. 1973. Flood routing methods for British rivers. *HRS Rep. INT 111.*

Price, R. K. 1973. Variable parameter diffusion method flood routing. *HRS Rep. INT 115.*

Thorn M. F. C. 1975. Monitoring silt movement in suspension in a tidal estuary. *Proc. XVI Congress I.A.H.R. Sao Paulo, July 1975.*

White, W. R. 1970. River gauging, *Channel 3* (*6*), 182-183.

White, W. R. 1971. Flat-vee weirs in alluvial channels. *Proc. ICE, 97 HY3*, 395-408.

White, W. R. 1971. A field comparison of the calibrations of a flat-vee weir and set of electro-magnetic flowmeters. *Wat. and Wat. Engng. 75* (*907*), 340-344.

White, W. R. 1971. The performance of two dimensional and flat-vee triangular profile weirs.

White, W. R. 1973. A mathematical model for the analysis of river diversions. *Proc ICE 55* (2), *371-389.*

White, W. R. 1971. The design and calibration of flat-vee weirs. *HRS Rep. INT 88.*

White, W. R. 1971. A field comparison of the calibrations of flat-vee weir and a set of electro-magnetic flowmeters. *HRS Rep. INT 94.*

White, W. R. 1972. The effect of adverse approach conditions on the calibrations of a compound weir. *HRS Rep. INT 114.*

White, W. R. 1972. Sediment transport in channels: A general function. *HRS Rep. INT 104.*

White, W. R. 1975. Field calibration of flow measuring structures. *Proc. ICE., 58* (*1*), 681-683.

White, W. R. and Ackers, P. A. 1973. A general function to describe the movement of sediment in channels. *Proc. XV Congress I.A.H.R. Istanbul, Turkey, 1973, 1*, 353-360.

White, W. R. and Crabbe, A. D. 1973. A comparison of observed and predicted sediment transport rates for some field and laboratory measurements. *HRS Rep. INT 122.*

White, W. R., Milli, H. and Crabbe, A. D. 1973. Sediment transport theories: a review. *Proc. ICE 59* (2), 265-292.

White, W. R., Milli, H. and Crabbe, A. D. 1974. Determination of total transport. *Proc. I.A.H.R. VIth Latin American Congress, Bogota, July 1974.*

White W. R., Milli, H. and Crabbe, A. D. 1973. Sediment transport: an appraisal of available methods. *HRS Rep. INT 119.*

White, W. R. and Price, R. K. 1975. The design of drainage channels using mathematical model techniques. *Proc. ICE 59* (2), 91-102.

White, W. R. and Whitehead, E. 1971. Field calibration of a compound weir at Manley Hall, River Dee. *HRS Rep. INT 101*.

White, W. R. and Whitehead, E. 1974. Field calibration of a compound weir at St John's Beck, Thirlmere. *HRS Rep. INT 135*.

White, W. R. and Whitehead, E. 1974. Field investigations of a triangular weir under drowned flow conditions. *HRS Rep. 137*.

1972. Trapezoidal drainage channels with lateral inflow. *HRS Rep. DE3*.

1970. Crump profile flat-vee weir. *HRS Rep. Ex 473*.

1970. Sediment movement in the approach channel to flat-vee weirs. *HRS Rep. Ex 476*.

1970. Triangular profile Crump weir: re-examination of discharge characteristics. *HRS Rep. Ex 477*.

1970. The triangular profile Crump weir. Effects of a bend in the approach channel. *HRS Rep. Ex 518*.

1972. Thames barrier: determination of silt shear strength. *HRS Rep. Ex 594*.

1973. Thames flood prevention investigation: continuous silt monitoring, 1970-73. *HRS Rep. Ex 671*.

TRANSPORT AND ROAD RESEARCH LABORATORY

Crowthorne, Berkshire, RB11 6AJ

The Transport and Road Research Laboratory carries out research into problems that arise in designing, building, maintaining and using public highways.

Within the Climate and Environment Section (now Environment Division) studies were completed to develop models for predicting urban and rural flooding within the United Kingdom in connection with storm water sewer and bridge or culvert design respectively.

Young, C. P. 1973. Urban drainage in the United Kingdom. *CIRIA Research Colloquium*, Bristol.

Young, C. P. and Prudhoe, J. 1973. The estimation of flood flows from natural catchments. *DOE, TRRL Report LR 565*.

WATER POLLUTION RESEARCH LABORATORY until 31 MARCH 1974

(now part of the WATER RESEARCH CENTRE; see page 90)

Elder Way, Stevenage SG1 1TH, Hertfordshire

The Laboratory is concerned with the effects of pollution on natural waters, including coastal waters, and with processes for purifying sewage and industrial wastes.

The sections studying the chemistry and biology of rivers are concerned among other things with the longitudinal and vertical mixing in freshwater streams, and with predicting times of passage of pollutants. Methods have been developed for measuring the flow of a stream with tracers using a lithium salt or radioactive bromine. The main function of the estuary and coastal pollution sections is the study of hydraulic mixing, dispersal, self-purification and diffusion; readily identifiable bacteria and radioactive isotopes are regularly used as tracers. The mathematics section plays an important part in all these studies.

The instrument section is developing permanent, portable, and submersible multivariable water quality monitors. It is concerned with automation at sewage works and with the development of counting and recording equipment for dispersion studies utilising radiochemical tracers.

Agg, A. R. and Wakeford, A. C. 1972. Field studies of jet dilution of sewage at sea outfalls. *Instn publ. Hlth Engrs J., 71*, 126.

Briggs, R., Agg, A. R. and Robertson, K. G. 1970. Counting and recording equipment for coastal and estuarine pollution studies. *Proc. Instn Electronic Radio Engrs Conf.*, 'Electronic Engineering in Ocean Technology', 15.

Downing, A. L. 1970. Water pollution research: past, present and future. *Chemy Ind., 286.*

Garland, J. H. N. and Hart, I. C. 1971. Trent research programme. Water quality relationships in the river system. *Inst. wat. Pollution Control Symposium*, Nottingham, April.

Jones, K., Briggs, R., Carr, J. G. and Potten, A. H. 1969. Automatic control of aeration in a fully nitrifying activated-sludge plant. *Instn publ. Hlth Engrs J., 68*, 271.

Knowles, G. 1970. Mathematical models as an aid in the solution of water pollution control problems. *Chemy Ind., 697.*

White, K. E. 1972. The use of radioisotopes for the measurement of flow and dispersion characteristics. *Inst. mech. Engrs. Conf. on Measurement Techniques in Air and Water Pollution*, London.

WATER RESEARCH ASSOCIATION

(until 31st March, 1974; see page 90),
Ferry Lane, Medmenham, Marlow,
Buckinghamshire

The Water Research Association served its members from the water supply industry and river authorities in applying research to a variety of biological, chemical, chemical engineering, civil engineering and physical fields. The Association received substantial grant aid from the Department of the Environment.

Anonymous. 1970 River flow measurement by dilution gauging. A manual of principles and practice using sodium dichromate as a tracer. *WRA tech. Paper, TP.74.* 85p.

Cole, J. A. 1970. Control rules for multi-use reservoirs and multi-reservoir systems. *Proc. 2nd Intl. Seminar for Hydrology Professors, Utah State Univ., Logan, Utah*, 343-378.

Cole, J. A. 1971. Optimal seasonal and short-term operation of a reservoir used for flood control and water supply. *Gentbrugge, International Association of Scientific Hydrology, International Symposium on Mathematical Models in Hydrology, Warsaw.* (In press).

Cole, J. A. 1971. Discussion of paper, by G. K. Young, G. T. Orlob and L. A. Roesner: Design criteria for using stochastic hydrology. *Proc. Am. Soc. Civ. Engrs J. Hydraul. Div., 97*, 583-585.

Cole, J. A. 1973. Examples of flow balance solutions for aquifer T and S by an automatic optimization procedure. *Int. Symp. on Development of Ground Water Resources, Madras, 1 (II-a)*, 13-27.

Cole, J. A. 1974. (Editor). Groundwater pollution in Europe. *Water Information Center, Inc., New York*, 600p.

Cole, J. A., Gunwardane, G. J. P. and Jones, Jennifer. Reservoir yield IV. Generalized curves of probability of emptiness of a reservoir, obtained from a two-season application of the Moran model. *WRA tech. Memo., TM. 60.* 44p.

Cole, J. A., Hirsch, H. and Bathurst, P. H. 1973. Calculating aquifer properties and baseflow compensation. Research lessons that have been drawn from an analogue and digital study of the Linstead Valley, Jamaica. *WRA tech. Paper, TP.96.* 58p.

Cole, J. A., Hirsch, H. and Thorn, B. M. 1972. Finding natural recharge patterns and abstraction phases to maximize yield of aquifers. *International Symposium on Mathematical Modelling Techniques in Water Resources Systems Environment Canada, Ottawa, 2,* 240-249.

Cole, J. A. and Sherriff, J. D. F. 1972. Some single- and multi-site models of rainfall within discrete time increments. *J. Hydrol., 17,* 97-113.

Davis, J. M. 1973. Groundwater modelling – a comparison between finite difference and finite element methods. *WRA tech. Memo., TM.87,* 9p.

Green, M. J. 1970. Calibration of the Brenig catchment and the initial effects of afforestation. *Gentbrugge, International Association of Scientific Hydrology, Publ. No. 96* (Symposium on the results of research on representative and experimental basins, Wellington, N.Z.), 329-345.

Green, M. J. 1970. Effects of exposure on the catch of rain gauges. *J. Hydrol. N.Z., 9,* 55-71.

Mawer, P. A., O'Kane, J. P. J., Sheriff, J. D. F., Wyatt, T. and Barrett, D. 1971. Some conjunctive desalting studies in the United Kingdom. *Wat. Resour. Bull., 7,* 831-843.

Mawer, P. A., Sherriff, J. D. F., Thorn, D. G. and Wyatt, T. 1971. Multiple resource studies. *WRA tech. Memo., TM.67.* 85p.

Mawer, P. A. and Thorn, D. G. 1973. Improved dynamic programming procedures and their practical application to water resources systems. *WRA tech. Memo., TM.77.* 29p.

Mawer, P. A. and Wyatt, T. 1974. Conjunctive desalination in water resource planning. Mathematical techniques for use in the economic assessment and operation of desalination in conjunction with conventional water resources, including case studies of three applications. *WRA tech. Paper, TP.108.* 97p.

Sherriff, J. D. F. 1970. Synthetic rainfall sequences. Mathematical analysis and synthesis of daily and hourly rainfall sequences for seven sub-catchments of the River Dee in North Wales. *WRA tech. Paper, TP.72.* 65p.

Taylor, C., Al-Mashidani, G. and Davis, J. M. 1974. A finite element approach to watershed runoff. *J. Hydrol., 21,* 231-246.

Thorn, D. G. 1972. A computer program for fitting seasonal probability distributions. *WRA tech. Paper, TP.92.* 23p.

Various authors. 1971. Artificial groundwater recharge conference. *WRA Proceedings Vols. I and II.* 742p.

Various authors. 1972. Groundwater pollution conference. *WRA Proceedings Vols. I, II and III.* 831p.

Various authors. 1973. Computer uses in water systems (including some papers on water resource topics). *WRA Proceedings Vols. I and II.* 471p.

WATER RESOURCES BOARD

(until 31st March 1974; see page 89)

Reading Bridge House, Reading, Berkshire

The Water Resources Board was, until 31st March 1974, the government agency responsible for advising the Secretary of State for the Environment and the Secretary of State for Wales on national policy in England and Wales; and assisting and advising river authorities in the conservation, redistribution and augmentation of water resources. Since 1st April 1974 the duties of the Board have been distributed amongst the Central Water Planning Unit, the Water Data Unit and the National Water Council, and much of the research work is being continued by the Water Research Centre.

The Board supported, initiated and undertook research over a wide range of subjects in its investigation of new ways of assessing water resources more accurately and of providing water economically. It began a desk study of possible storage in the Wash, a study now being continued by the Central Water Planning Unit. Other studies include instrumentation, telemetry and data processing, resource allocation, desalination, fisheries, ecology, economics, groundwater development and artificial recharge and operational research. Many of these have been reported upon and most are being continued by the new central organisations. Some of the more important hydrological work undertaken by the Board is outlined below.

Data Collection and Processing The Board was responsible for detailed discussions of hydrometric networks with the 29 river authorities; these networks included river gauges, climate and rainfall stations and water quality stations. Comprehensive networks were approved and programmes of implementation agreed. Similarly groundwater observation well networks were discussed and agreed for all major aquifers within each river authority. Good progress was made in the development of these basic hydrological data collection systems.

The processing of data from these systems has involved the storing, updating and retrieval of all available historical records on computer magnetic tape. This has allowed the preparation of 'The Surface Water Year Book of Great Britain' and the 'Groundwater Year Book' to be produced in a new format, utilising computer print-out of information and statistics. The Surface Water Year Book for 1966-70 inclusive and the Groundwater Year Book for 1968-70 have been produced, comprising comprehensive indexes of measuring stations, graphical and tabulated records from selected stations and descriptions of the supporting computer retrieval services.

River Dee Regulation Research Programme This study was designed to enable river regulation systems to be operated more efficiently for the diverse interests of water supply, flood alleviation, land drainage and fisheries. A telemetry system was installed and has provided operational experience and data for the application of alternative control strategies. A dense rainfall network and radar system was installed to determine if rainfall amounts could be forecast to help in the management of a multi-purpose reservoir system.

Generation of Synthetic Hydrological Data Limited records of river flows have been extended by the use of statistical techniques to provide valid long-period records.

Instrumentation The Board collaborated with industry and river authorities to develop reliable proven equipment, including climatological monitoring stations, automatic rain gauges, river quality monitoring stations and fish counters.

River Gauging The Board collaborated with Atomic Energy Research Establishment, Harwell, and several river authorities to successfully develop a method of flow measurement using ultrasonic techniques. A prototype system has been in operation on the River Thames since 1972.

Hydrological Design of Water Resource Systems Complex water resource systems have been modelled using two stage simulation techniques, the first to generate synthetic river flows at various points within the system and the second using these synthetic flow data in simulation of the reservoir system itself. A simulation model has been successfully developed and operated to study several rivers, reservoirs, several aquifers and estuary storage in eastern England.

Groundwater Hydrology

Artificial Recharge The Board gave high priority to the investigation of artificial recharge both as a means of affording partial treatment and to augment groundwater storage. Field experiments have been carried out using spray irrigation, basin, and well recharge techniques at sites on the Triassic sandstone north of Nottingham and a report describing the results has been published. Further studies and reports have examined the hydrogeological and economic feasibility of artificial recharge of the Chalk and Tertiary Sands in the London Basin and as a result a first stage development is currently being designed for the lower Lee Valley.

Combined Use The principle of utilising groundwater storage together with surface sources in order to provide an increased yield for the system as a whole has been extensively investigated by the Board. Studies have involved detailed investigation of both surface water and groundwater hydrology in the areas concerned and the establishment of the factors controlling the interactions between aquifers and surface flows. Pilot studies have been carried out in many areas involving both the Chalk and Triassic sandstone aquifers and a series of reports produced.

Mathematical Models The Board developed a large number of models of groundwater and surface water systems to predict the effects of both artificial recharge and combined use developments. Such models have played a leading role in the interpretation of the hydrogeological and hydrological regimes and in the design of development proposals. In addition to the application of modelling techniques to large scale problems, studies have also been carried out in the field of well hydraulics. References to mathematical models are made in many of the Board's publications in addition to specific modelling reports.

Geophysical Techniques Extensive use of a wide range of geophysical logging techniques has been made in the field of groundwater hydrology and such methods have led to a detailed understanding of the mechanisms of groundwater flow in the aquifers of the United Kingdom.

Isotope Tracer Techniques The use of tracers to determine the rate and direction of groundwater flow has been studied. Tracers, particularly tritium, have additionally been used to estimate rates of flow in the unsaturated zone, and to establish the degree of hydraulic connection between rivers and aquifers.

Bloomer, R. J. G. and Sexton, J. R. 1971. The problems encountered in synthetic river flow generation procedures. *Int. Symp. on Mathematical Hydrology.*

Brereton, R. 1972. A reappraisal of the $^{40}AR/^{39}AR$ stepwise degassing technique. *Geophyscial J. roy astron. Soc., 27,* (5), 449-478.

Bussell, R. B., Harrold, T. and Grinstead, W. 1972. The Dee weather radar project. *World Meteorological Organisation Symp., Geilo, Norway.*

Downing, R. A., Backshall, W. F. and Law, R. M. 1972. Great Ouse groundwater study. *Wat. & Wat. Eng., 76,* 215-223.

Downing, R. A. 1971. Groundwater management in England and Wales. *Quart. J. eng. Geol., 4.*

Downing, R. A., Davies, Mrs M. C., Pontin, J. M. A. and Young, C. P. 1972. Artificial recharge on the London Basin. *Hydrol. Sci. Bull.,* (2), *17,* 183-187.

Downing, R. A., Williams, B. P. J. and Lovelock, P. E. R. 1972. Aquifer properties of the Bunter sandstone in Nottinghamshire, England. *Proc. 24th international geological Congress, Montreal.*

Edmunds, W. M., Taylor, B. J. and Downing, R. A. 1970. Mineral and thermal waters of the United Kingdom. *XXIII international geological Congress, 18,* 139-158.

Herschy, R. W. 1970. The magnitude of errors at flow measurement stations. *UNESCO Symposium on Hydrometry. Koblenz.*

Herschy, R. W. 1971. The use of weirs and flumes on stream gauging. *World Meteorological Organisation.* Report of a Working Group of the Commission for Hydrology. Geneva. Technical Note No. 117. WMO No. 280.

Ineson, J. 1970. The development of groundwater resources in the United Kingdom. *Instn. Water Eng., 24,* (3), 155-177.

Jamieson, D. G. and Onstud, C. A. 1970. Modelling the effect of land use modification on run-off. *Wat. Resour. Res., 6,* (5), 1287-1295.

Jamieson, D. G. 1972. Dee Research Programme (1): Operating multi-purpose reservoir systems for water supply and flood alleviation. *Wat. Resour. Res., 8,* (4).

Jamieson, D. G. and Wilkinson, J. C. 1972. Dee Research Programme (3): A short-term control strategy for multi-purpose reservoir systems. *Wat. Resour. Res., 8,* (4).

Jamieson, D. G., Smith, D. K. and Wilkinson, J. C. 1973. Evaluation of short-term operational policies for a multi-purpose reservoir system. *Proc. Inter. Fed. of Automatic Control Symp. on the Control of Water Resources Systems, Haifa, Israel.*

Jamieson, D. G., Wilkinson, J. C. and Innit, R. P. 1972. Hydrological forecasting with sequential deterministic and stochoastic stages. *Proc. Inter. Symp. on Uncertainties in Hydrological and Water Resource Systems, Tucson, USA.*

Monkhouse, R. A. 1975. An assessment of the groundwater resources of the Lower Greensand in the Cambridge-Bedford region. *Water Resources Board.*

Naylor, J. A. (In press). The groundwater resources of the River Gravels of the Middle Thames Valley. *Water Resources Board.*

Newsome, D. H. 1970. Re-use of water in the context of water resources development. *The Chem. Eng.*, (235).

Newsome, D. H. and Jackson, S. Development in water reclamation and future supplies. *Symp. on Management of Water in the Iron and Steel Industry.* The Iron and Steel Institute May 1970.

Newsome, D. H., Bowden, K. and Green, J. A. 1971. A mathematical model of the Trent River system. *Symp. on the Trent research programme at the University of Nottingham.* The Institute of Water Pollution Control, Paper No. 5.

Newsome, D. H. 1972. The River Trent – a guide to management. *Symp. on Mathematical Modelling Techniques in Water Resources Systems, Ottawa, Proceedings, 3,* 613-627.

Newsome, D. H., Bowden, K. and Green, J. A. 1972. Trent mathematical model (1) Construction. *Symp. on Advanced Techniques in River Basin Management, Birmingham. Proceedings,* 107-114.

Oaks, B. D. and Ponton, J. M. A. (In press). Modelling of groundwater and surface water systems II – A digital model of the chalk of the Lambourn Valley, Berkshire: Simulation of groundwater development to augment river flow. *Water Research Centre.*

O'Neill, P. G. 1972. A mathematical programming model for planning a regional water resource system. *J. Int. wat. Eng., 27,* (1), 47-61.

Phillips, S. and Lambert, N. K. (In press). Forecasting groundwater levels by means of time series analysis. *Water Research Centre.*

Pontin, J. M. A. and Boulton, N. S. 1971. An extended theory of delayed yield from storage applied to pumping tests in unconfined anistropic aquifers. *J. Hydrol., 14,* 53-65.

Radford, P. J. and Greenwood, D. J. 1970. The simulation of gaseous diffusion in soils. *J. Soil. Sci., 21,* (2).

Richards, H. J. Groundwater resources in northern England. Geological aspects of development and planning in northern England. Editor, Warren, P. T. *York. Geol. Soc., 1970,* 14-21.

Richards, H. J. 1972. River basin groundwater development – principles and the national picture. *Symp., Instn. civ. Engrs, November 1972.*

Richards, H. J. 1971. Development, protection and replenishment of groundwater resources. *Economic Commission for Europe, Seminar on Selected Water Problems in Southern Europe, Zagreb.*

Richards, H. J. and Smith, D. B. 1971. Selected environmental studies using radioactive isotopes. *Fourth United Nations Conference on the Peaceful Uses of Atomic Energy, Geneva.*

Richards, H. J. and Skinner, A. C. (In press). Development of groundwater in Triassic Sandstones of the Fylde for combined use with surface resources. *Water Resources Board.*

Rydz, B. 1971. Water resources for industry. Society of Chemical Industry. *Symp. on Use and Re-Use of Water in Industry.*

Rydz, B. 1971. Regional water resources analysis. *Proc. Instn. civ. Eng. 49,* 129-145.

Satchell, R. L. H. and Edworthy, K. J. 1972. Artificial recharge: Bunter Sandstone. Trent Research Programme, Vol. 7. *Water Resources Board.*

Satchell, R. L. H. and Wilkinson, W. B. 1973. Artificial recharge in the United Kingdom with special reference to the London Basin. *United States Geological Survey. Inter. Symp. on Underground Management and Artificial Recharge.*

Satchell, R. L. H. and Edworthy, K. J. 1971. Recharge of Nottinghamshire Bunter Sandstone. *Institution of Water Engineers, Symp. on Advanced Techniques in River Basin Management, Birmingham. Pro*ceedings, 55-73.

Sexton, J. R. and Jamieson, D. G. 1973. Improved techniques for water resource systems design. *Proc. of WMO, IAHS, UNESCO Symp. on Design of Water Resource Systems with Inadequate Data, Madrid.*

Sharp, R. G. and Harding, J. 1970. Hydrometeorological data processing in the United Kingdom. *World Meteorological Organisation Technical Conference, Geneva.*

Sharp, R. G. and Watson, D. 1971. Correlation of river flow data for yield assessment. *Paper to Hydrological Group, I.C.E. Proceedings, Informal Discussions.*

Taylor, L. E. 1971. Water resources in the South East. *Proc. Symp. Land Use and Landscape Quality, University of Reading.*

Taylor, L. E. 1973. Storing fresh water in the Wash. *The Central No. 146.* Spring 18-20.

Warne, A. E. 1973. The Trent mathematical model. *Proc. Symp. on the Use of Mathematical Models in Water Pollution Control,* Newcastle upon Tyne.

Warne, A. E., Bewin, D. B., Chang, M. and Porter, K. 1972. Trent Mathematical Model (2) Development. *Instn. of Wat. Engrs. Symp. on Advance Techniques in River Basin Management, Birmingham.* Proceedings, 115-128.

Water Resources Board. (In press). Groundwater Resources of the Vale of York. *Water Resources Board.*

Water Resources Board. (In press). Artificial recharge on the London Basin – III Economic and engineering desk studies. *Water Resources Board.*

Wilkinson, W. B. 1973. The development of groundwater resources – Past, present and future. *J. Waterworks Officers Ass., 10* (1), 16-20.

Wilkinson, W. B. and Oakes, D. B. 1973. Use of digital models in the development of groundwater resources. Informal discussion, Hydrological Group. *Proc. Instn civ. Engrs, Part 2.*

Wright, C. E. 1971. Catchment characteristics influencing low flows. *Wat. & Wat. Eng., 74,* 468-471.

Wright, C. E. 1975. Combined use of surface and groundwater in the Ely Ouse and Nar catchments. *Water Resources Board.*

FORESTRY COMMISSION
RESEARCH AND DEVELOPMENT DIVISION
Alice Holt Lodge, Farnham, Surrey
and
Roslin, Midlothian

The Forestry Commission acts as the national forest authority and as the national forest enterprise. The functions of the forest authority are to advance knowledge and understanding of forestry and trees in the countryside, to develop and ensure the best use of the country's timber resources and to promote efficiency and development in the home timber industry, to undertake research relevant to the needs of forestry, to combat forest and tree pests and diseases and to initiate Plant Health Orders when appropriate.

The functions of the Commission as the forest enterprise are to develop forestry and increase the production of wood for industry by the extension and improvement of the forest estate, to protect and enhance the environment, to provide for recreation, to stimulate and support, by forestry, the local economy in areas of depopulation, to further the integration of forestry and agriculture and in pursuit of these objectives to manage the estate as profitably as possible. The R & D Division is working on projects concerned with water relations of forest crops and moisture characteristics of forest soils at both its stations. The Division collaborates with the Institute of Hydrology in forest catchment research, mainly by provision of sites and facilities.

The Site Studies (S) Branch at Alice Holt is working on problems concerned with the growth of trees on a range of soil types; changes in water regime following afforestation; drainage, cultivation and fertilisation of former heathland; land use problems such as the rehabilitation of opencast coal tips, gravel workings and recreation sites; the effect of trees on site conditions, especially differences between species; the growth of trees on chalk soils, their moisture abstraction pattern and nutrition.

The Site Studies (N) Branch at Roslin is working on the relationships between moisture, aeration and root growth in a range of soil types (mainly clayey soils such as surface water gleys and peaty gleys, ironpan soils and deep peats) and the effect of afforestation on the water balance and soil conditions. Field studies are supplemented by laboratory work on moisture retention. Studies are being made of the shrinkage and other irreversible changes which follow tree growth on peat soils.

The Physiology Branch at Roslin is investigating the effects of excess soil moisture on the physiology and development of the root systems of important commercial species.

Binns, W. O. 1969. Water use by tree plantations. *J. Hydrol., 7,* 109-110.

Binns, W. O. 1970. Forests, rainfall and run-off. In: *Association of River Authorities Yearbook 1969*, 19-23.

Binns, W. O. 1970. Some effects of tree-growth on deep peat. *Proc. 3rd int. Peat Congress, Quebec 1968*, 358-365.

Darlington, E. 1971. A simple water-level dipstick for boreholes. *J. agric. Engng Res. 16*, 423-424.

Fourt, D. F. 1968. Sitka spruce, shelter and moisture. *Res. Dev.* Pap. *For. Commn, 72*.

Fourt, D. F. and Hinson, W. H. 1970. Water relations of tree crops: a comparison between Corsican pine and Douglas fir in south-east England. *J. appl. Ecol., 7*, 295-309.

Hinson, W. H. 1971. A simple method of obtaining an analogue output from capacitance as a transducer. *J. phys. E: Scient. Instrum. 4*, 778-779.

Hinson, W. H. and Kitching, R. 1964. A readily constructed transistorised instrument for electrical resistance measurements in biological research. *J. appl. Ecol., 1*, 301-305.

Kitching, R. 1965. A precision portable electrical resistance bridge incorporating a centre zero null detector. *J. agric. Engng Res., 10*, 264-266.

Kitching, R. 1967. Water use by tree plantations. *J. Hydrol., 5*, 206-213.

HARWELL (Research Laboratory of the UK Atomic Energy Authority)

AERE, Harwell, Oxfordshire, OX11 0RA

Harwell undertakes materials research and development for the nuclear power programme, applied research in collaboration with industry and a supporting programme of underlying research. Its contribution to hydrological research includes isotope applications and desalination research.

A low-level tritium and Carbon-14 measurement laboratory studies applications to groundwater recharge and movement, in close collaboration with NERC and the Water Research Centre. Movement of water through the unsaturated zone of chalk and sandstone is being studied using natural tritium. The water in confined aquifers is being dated using Carbon-14 supported by stable isotope measurements of deuterium and Oxygen-18. Radioactive tracers are used in other groundwater velocity and pollution studies. Tritium is used for evaluating water movement in or beneath mine tip areas and in studies of unstable hillsides. River flow and flood flow measurements have been made using radioactive tracers as a method of field calibration of weirs and structures.

The Atomic Energy Authority in conjunction with industry, is undertaking research into desalination by the process of distillation, freezing, reverse osmosis and electrodialysis. The last two processes are particularly relevant to the water resources of the United Kingdom.

Smith, D. B., Wearn, P. L., Richards, H. J. and Rowe, P. C. 1970. Water movement in the unsaturated zone of high and low permeability strata by measuring natural tritium. *Isotope Hydrology 1970, 73, Proc. Symp., IAEA, Vienna.*

Smith, D. B., Wearn, P. L. and Parsons, T. V. 1968. Accuracy of open channel flow measurements using radioactive tracers. *Rep. AERE-R 5676, HMSO.*

UKAEA. 1967. Desalination and its role in water supply. HMSO.

Hydrological Tracing using Isotopes. 1974. (A Harwell industrial research brochure available from Mr D. B. Smith, Harwell, Oxfordshire, OX11 0RA.)

LABORATORY OF THE GOVERNMENT CHEMIST

Cornwall House, Stamford Street, London, SE1 9NQ

The Laboratory of the Government Chemist provides independent and impartial chemical analytical, advisory and investigatory services to Government Departments and several statutory bodies.

The Radiochemical Sub-division of the Laboratory acts for the Radiochemical Inspectorate of the Department of the Environment and for the Scottish Development Department by providing analytical facilities for the determination of radionuclides in potable and effluent water. In addition, there is a continuing programme of improvement and development of radiochemical methods of analysis for use in this work.

The Water and Waste Water Chemistry Sub-division of the Laboratory undertakes analyses and advises on the quality of potable and industrial waters on behalf of Government Departments, in particular, the Property Services Agency of the Department of the Environment; the chemical analysis of rainwater is also a continuing service provided for the Meteorological Office. It plays a part, both in committees and through collaborative practical studies, in the work of national and international bodies concerned with the development of standard and reference methods of analysis.

Special and continuing interests have included research work on fluoridation and the biodegradability of anionic and non-ionic detergents involving the development of specific analytical methods for these detergents and their degradation products. More recently, de-icing and anti-icing chemical agents and their breakdown products have been studied and methods developed for determining their concentration in run-off waters. Current research interest concerns the development of methods of analysis for the determination of a wide variety of trace metallic and organic pollutants in water and liquid effluents in the microgram per litre range.

Darrall, K. G., Hammond, G. C. M. and Tyler, J. F. C. 1973. The determination of plutonium-241 in effluents. *Analyst, 98,* 358.

Darrall, K. G., Richardson, P. J. and Tyler, J. F. C., 1973. An emanation method for determining radium using liquid scintillation counting. *Analyst, 98,* 610.

Dennis, A., Evans, W. H. and Patterson, S. J. 1972. Spectrophotometric determination of glycols. *Water Treatment and Examination, 21* (4), 350.

Evans, W. H., David, E. J. and Patterson, S. J. 1973. Biodegradation of urea in river waters under controlled laboratory conditions. *Water Research, 7,* 975.

Evans, W. H. and Dennis, A. 1973. Spectrophotometric determination of low levels of mono-, di- and triethylene glycols in surface waters. *Analyst, 98,* 782.

Evans, W. H. and David, E. J. 1974. Biodegradation of mono-, di- and triethylene glycols in river waters under controlled laboratory conditions. *Water Research, 8,* 97.

Evans, W. H. and Partridge, B. F. 1974. Determination of ammonia levels in water and waste water with an ammonia probe. *Analyst, 99,* 367.

Evans, W. H. and Sayers, G. S. 1972. Visual estimation and spectrophotometric determination of zinc in potable waters with 4– (2′ thiazolylazo) resorcinol. *Analyst, 97,* 453.

Evans, W. H. and Stevens, J. G. 1972. An investigation into the determination of nitrate in potable waters and effluents by the reduction method employing Devarda's alloy. *Water Pollution Control, 71* (1), 98.

Patterson, S. J., Scott, C. C. and Tucker, K. B. E. 1970. Nonionic detergent degradation, Part III: Initial mechanism of the degradation. *J. Amer. Oil Chemists' Soc., 47,* 37.

RIVER AUTHORITIES (ENGLAND AND WALES)

Until 31st March, 1974, there were 29 River Authorities including the Thames Conservancy and the Lee Conservancy Catchment Board; they covered the whole of England and Wales, except part of the London area. They were created as the result of the Water Resources Act of 1963 and were responsible for the control, management and development of water resources, and the functions of the former river boards; land drainage, freshwater fisheries, prevention of river pollution and, in some cases, navigation.

They were required under the 1965 Act to introduce, each in its own area, hydrometric schemes to produce data on a wide range of hydrological variables. Many river authorities carried out hydrological research, which is now continuing under the new Regional Water Authorities (see page 94).

WATER SUPPLY UNDERTAKINGS

The water supply undertakings (absorbed into Regional Water Authorities on 1st April, 1974) were required to measure the amounts taken by them for the need of communities rather than the total surfaces or underground water resources from which they derive their supplies. As regards residual surface water resources, the undertakings usually established that a prescribed flow had been left in the stream from which abstraction had taken place. Their abstractions were licensed by the local river authority under the Water Resources Act, 1963.

A few undertakings utilising surface water resources voluntarily made additional measurements in order to determine over any period of time the amount of water draining from their catchments. In such cases the data obtained were supplied to the Water Resources Board: in England and Wales through the local river authority and in Scotland through the Scottish Development Department.

SCOTLAND

SCOTTISH DEVELOPMENT DEPARTMENT

The Scottish Development Department is responsible for the conservation of water resources and has prime responsibility for coordinating the establishment of a national network of river gauging stations. The operation and extension of this network is carried out by SDD itself and by some of the Scottish river purification boards. Although activity is orientated more towards data collection than fundamental research, investigations are carried out into practical problems associated with the development and operation of water level recording instrumentation, current meter calibration, stage/discharge relations in loose boundary open channels, errors in river gauging and other associated fields.

The Department cooperates with Universities and other authorities on schemes with a hydrological interest such as flood alleviation and control, flood warning telemetry, estuary surveys, effects of pollution on rivers and lochs and the monitoring of migratory fish.

The Water Data Unit of the Department of the Environment provides processing facilities for river flow data to SDD, the RPBs and other authorities. This data is described in the joint SDD/DOE publication 'The Surface Water Year Book of Great Britain'.

RIVER PURIFICATION BOARDS

Until 15 May 1975 there were nine Scottish river purification boards covering the mainland south and east of the Caledonian Canal. A restructuring of these boards became effective on 16 May 1975 when seven boards were formed to cover the whole of the mainland. As part of their function these boards undertake flow measurement in rivers and small streams, rainfall and other climatic measurements, analyses of hydrological data and hydrological and hydrographical surveys.

NORTHERN IRELAND DEPARTMENT OF AGRICULTURE

FORESTRY DIVISION

Dundonald House, Upper Newtownards Road, Belfast BT4 3SB

Runoff from a catchment

Runoff data have been analysed for 12 years (1960-71) recordings from the Woodburn catchment near Belfast. Although results were inconclusive the experiment is being continued by the Northern Ireland Department of the Environment.

Savill, P. S. and Weatherup, S. T. C. 1974. The effect of afforestation on water runoff in the Woodburn catchment area. *Forestry XLVII*, 45-56.

Artificial drainage of peats and gleyed soils

Experiments have been established to ascertain the effects of different drainage intensities on the level of water in boreholes and on tree growth and stability.

Savill, P. S., Dickson, D. A. and Wilson, W. T. 1974. Effects of ploughing and drainage on growth and root development of Sitka spruce on deep peat in Northern Ireland. *Proc. Int. Symposium on Forest Drainage*, 241-252.

Ploughing in relation to tree growth and stability

Shallow plough furrows can restrict tree root development on wet soils and increase windthrow risk. Experiments are being undertaken to devise alternative methods of ground preparation for planting.

THE DEPARTMENT OF THE ENVIRONMENT FOR NORTHERN IRELAND

THE WATER SERVICE

Stormont, Belfast BT4 3SS

The Department is responsible for the management of all aspects of water in Northern Ireland excluding arterial and urban drainage and excluding certain provisions for water recreation. The Departmental emphasis is on water conservation, water supplies and sewage disposal. The enabling legislation is contained in the Water Act (NI) 1972, the Water and Sewerage Services (NI) Order 1973 and the Water Regulations (NI) 1974. The Department is advised by the Northern Ireland Water Council.

The Department prepares regional reports on water supplies and water quality. A draft Water Management Programme for Lough Neagh has been prepared in accordance with the Water Act (NI) 1972.

The Department's hydrometric, meteorological, water quality and hydrogeological monitoring networks are being expanded to improve the data on which future water resource management will be based.

Although the Departmental emphasis is on its responsibilities as the Water Service for Northern Ireland, the Department actively supports water research by means of staff involvement and limited financial assistance. Much of this research supported directly or indirectly concerns hydrology.

NORTH OF SCOTLAND HYDRO-ELECTRIC BOARD

Records of rainfall and run-off from a small catchment in Argyll have been collected for a period of 23 years. These have been analysed with particular reference to the catchment water balance and evaporation estimates have been compared with observed losses from the catchment over long and short term periods.

Comparisons between standard daily rain gauges and gauges exposed at ground level, surrounded by a four foot square plastic grating as devised by the Institute of Hydrology, have been made at three sites in the Scottish Highlands for two years. Results show that in two instances there is virtually no difference between the catches of the two types of gauges; these sites are those typical of a Scottish heath exposure, the gauges being surrounded by rough grass, peat hags, heather and low scrub, but in the third case results repeating those typical of an English climatological station were obtained, i.e. the ground level gauge catches 5 to 6 per cent more rainfall than the standard gauge. Here the gauges were exposed on the flat surface of an old tunnel spoil tip which is now covered with moss and low grass and its surface must be aerodynamically similar to a lawn.

Trials with various types of transmitting raingauges are in progress. Radio or land-line telemetry links the tipping bucket rain gauge to the nearest power station. Results so far have given useful information but the equipment is not yet sufficiently trustworthy for full reliance to be placed upon it. Nine raingauges have been operated successfully at times but not continuously. One of the most troublesome features of the radio telemetry gauges has been the reception of spurious signals caused apparently by atmospherics or man-made transmissions not intended for our receiver. So far it had proved impossible successfully to relate the rainfall recorded via the radio with the catches of the check gauge.

Land-line telemetry had been more successful but here too snags have been encountered and close correlation with the check gauge is often not achieved.

Tipping bucket raingauges and paper tape data loggers similar to those used to record traffic counts are to be installed on one northern catchment to provide real time indication of rainfall intensities over a network in the catchment enabling storms to be flood-routed from initial input to provide flood peaks and runoff hydrographs to the power station.

Jarvis, R. M. 1975. Flood Assessment for Hydro-Electric Project, BNCOLD Symposium, Newcastle, 1975.

Reynolds, G. 1969. Rainfall runoff and evaporation on a catchment in West Scotland. *Weather, 24,* 90-98.

Reynolds, G. 1971. Ground level raingauges in Scotland. *Weather 26*, 379-381.

Reynolds, G. 1975. Extreme rainfall estimation for flood studies in the Scottish Highlands. BNCOLD Symposium, Newcastle, 1975.

WATER REORGANISATION IN ENGLAND AND WALES

Fundamental reorganisation of the water industry took place in April 1974. 29 river authorities, 157 water undertakings and 1393 sewerage and sewage disposal authorities were replaced by 10 new multi-purpose authorities (Regional Water Authorities) with responsibilities for water conservation and supply, sewerage and sewage disposal, the control of water pollution, land drainage and flood prevention, recreation, fisheries and some inland navigation. The labour force employed on these services is about 75,000. The philosophy behind this change was to rectify the administrative weakness inherent in the existing separation between quantity and quality and the difficulty of securing the effective execution of integrated proposals through so many different organisations.

The organisation of the regional water authorities reflects fairly closely the recommendations in the Ogden report 'The New Water Industry – Management and Structure'. Corporate Management Teams comprise generally the chief executive and directors of finance, operations, resource planning, scientific services and administration. The main variation in some authorities is in the placing of responsibility for scientific services, fisheries, recreation, public relations and personnel.

The changes in the executive responsibility for national water have also been complemented by changes in central Government support for research and planning to produce a greater concentration of effort and to introduce machinery better adapted to serve the new industry; they also came into force on 1st April 1974.

The Water Resources Board, which was set up in 1963 with responsibility to advise Government on water conservation, has been abolished (see page 78). During its ten-year life the Board showed that future demands for water could be met by the development of a limited number of large new water sources and the transmission of water over comparatively large distances using the existing river systems. The success of this strategy depends upon the maintenance and improvement of the quality of river water. The planning, control and carrying out of works to improve water quality can only be achieved in relation to particular river basins taking account of the natural characteristics of the river (and all rivers are different), of the effluents discharged into the river and of the local works needed to secure the designed quality standards.

Water authorities are required to carry out a survey of water in their area, to estimate the future demand for water and to prepare a plan to cover a period of 20 years, with periodic seven-year revisions.

NATIONAL WATER COUNCIL

1, Queen Anne's Gate, London SW1 9BT

The Water Act 1973 lays down that the national policy for water is the responsibility of the Secretary of State for the Environment and the Minister of Agriculture, Fisheries and Food. However, given the creation of the new multi-purpose Water Authorities, the Government considered that there was no need to interpose a national body between the authorities and Ministers. The need for a body like the Water Resources Board arose from the existence of a much larger number of authorities concerned with water services. They did consider, however, that there was a need for a consultative and advisory body at national level and they have set up a statutory *National Water Council* for this purpose. The Council consists of all of the Chairmen of the Water Authorities and of others nominated by the Government. They draw together advice on matters of common interest and act as a link between the Government and the authorities. Where appropriate they provide services for the regional authorities on a collective basis. The Council's other duties include assisting the water authorities in their preparation of plans and with research activities, and the establishment of schemes for the testing and approval of water fittings and for training and education.

WATER RESEARCH CENTRE

Medmenham Laboratory
Henley Road, Medmenham, Marlow, Bucks SL7 2HD
Stevenage Laboratory
Elder Way, Stevenage, Herts SG1 1TH

The Medmenham Laboratory covers all aspects of water supply, ranging from the raw water resource through all stages of treatment, distribution and use.

The Stevenage Laboratory is concerned with processes for purifying sewage and industrial wastes and with the effects of pollution on rivers and coastal waters.

Hydrological topics included in the Centre's programme of work are as follows:

Reservoirs

Evaluation of water yields in terms of reliability of supply, for given storage capacity and input hydrology. For the design and operation of complex systems, such as interconnected reservoirs, pumped storage and river regulating schemes, hydrological and operational research principles are jointly brought to bear.

The limnology of stored water, in terms of thermal

stratification, chemical balances and algal production, is studied, with a view to improved management of reservoir water quality.

Tracer studies of retention times in bankside storage reservoirs.

Groundwater

Studies to aid optimal management of existing and development of new groundwater resources. These include surface and downhole geophysical methods, borehole drilling and sampling, well pumping tests, chemical and hydraulic properties of rock cores, petrological examination of aquifer materials.

Investigations with the use of artificial recharge as a means of water conservation and treatment. Pilot studies in the Chalk, Bunter Sandstone, Folkestone Sands and Superficial deposits.

Determination of factors likely to cause pollution of groundwater sources, in particular, hazardous waste in landfill sites and nitrates. Detailed hydrogeological and chemical examination of up to ten existing landfill sites, mathematical modelling and long term monitoring. Field studies of the concentration of nitrates in the unsaturated and saturated zones of the principal aquifers in the UK and predictions as to how the concentration will change in time. Study of the pollution of groundwater systems resulting from mine workings and wastes and methods of aquifer rehabilitation.

Semi-automatic systems for the analysis of well pumping tests and studies of loss of efficiency in wells and methods of rehabilitation. Well screen and pack design using large model simulators.

Digital and analogue modelling of groundwater systems to predict movement of water and pollutants. Joint groundwater surface water modelling studies.

Rivers

Chemical and biological studies, with particular emphasis on the effects of discharges of sewage effluents and stormwater on the biota and water quality of rivers.

Hydrometric developments in rivers, sewers and water mains, including ultrasonic and electromagnetic gauging stations and dilution gauging techniques. Radiochemical studies of flow, retention time and dispersion phenomena.

Time series analysis to interpret water quantity, quality and flow variations.

Real time control of a regulated river system, with telemetry of radar-based rainfalls, river flows and reservoir contents to a flow forecasting centre. Demonstration project of the Welsh Dee.

Estuaries, Coasts

Hydraulic mixing, dispersal, self-purification and diffusion studies.

Multisource Studies

Studies to link surface and groundwater resources and use of desalination plants and industrial reclamation.

Instrumentation

Evaluation of commercially available sensors and systems for maintaining water quality. Development of instrumental system for river management and control of treatment processes; the work includes development of more reliable sensors for management of DO, turbidity, TOC, ammonia, etc., development of telemetry systems and computerised control stations. General advisory and backing service to research divisions and members.

Geophysical service for boreholes.

Development of heat pulse flowmeter to log vertical flows in boreholes.

THE CENTRAL WATER PLANNING UNIT

Reading Bridge House, Reading, Berks RG1 8PS

This Unit came into being on 1st April 1974 and is based on a nucleus of the former Water Resources Board's staff strengthened by the addition of disciplines concerned with water quality and economic analysis. Its principal function will be to keep under review national and strategic aspects of water services planning and operation in England and Wales particularly where these affect more than one water authority; to identify problems likely to arise in connection with possible long-term developments and carry out or promote related studies, including the development of new planning techniques. The Unit thus has an independent role in providing a service of professional assessments to the National Water Council, Government Departments and the new water authorities. The Unit's programme of work is determined by a steering committee which is presided over by the Chairman of the National Water Council with members drawn from the Council of the Water Research Centre (see above) and Government Departments.

The following are a selection of the projects which feature in the Unit's hydrological work.

Wash Feasibility Study

This study was authorised in 1971, begun by the Water Resources Board and is intended to determine the feasibility of building bunded freshwater reservoirs on the southern foreshore of the Wash. Catchments draining into the Wash have been studied to determine variations in quantity and quality of streamflow.

Severn to Thames Water Transfers

Studies have been made of the water quality and ecological problems involved in transfer of water from the river Severn to the river Thames, in addition to studies of the operation of multiple sources regulating these rivers.

Reliability Studies

To examine the effects of changing design criteria by review and appraisal of ways in which operational practice modifies the hydrological reliability of supplies. and economic importance of supplies to different users.

Multi-Site Flow Data Generation

Appraise existing techniques, to improve these and develop methods to produce compatible synthetic data for gauging stations having low natural correlations and preserve the characteristics of the historic records.

Re-assessment of Multi-source Yields

Extension of short-period flow records by correlation with historical weather information records held by the Meteorological Office to help increase data available for yield assessment.

Optimum Development of Combined Sources

To develop methods to help determine the most efficient way of using multiple resource systems of surface and groundwater.

Groundwater Quality
A comparative analysis of information on nitrate concentrations in groundwater to determine any possible effects on regional or national use of groundwater.

River Regulation Losses
With water authorities to examine river regulation schemes to determine losses and hence provide more efficient control.

Water Atlas of the United Kingdom
To compile with the Water Data Unit a series of selected maps.

Dee Weather Radar and Hydrological Forecasting
The radar study also involves Meteorological Office, Royal Radar Establishment and Plessey Radar and has produced a system which converts rainfall information for transmission as digital information or displayed on a RV screen. The Hydrological Forecasting study is funded by CWPU and WRC and carried out by IH to help control river regulation on the river Dee.

Artificial Recharge
Appraise the role of artificial recharge of aquifers in water resource development and help with funds and specialist studies of early stage programmes.

Pollution of Aquifers by Seawater
Determine the problems of saline intrusion as they relate to development and fund necessary research directed towards improved operation of aquifers.

Birtles, A. B. and Nutbrown, D. A. 1975. Some water resource system studies in the UK. *Int. Symp. on Application of Mathematical Models in Hydrology and Water Resources Systems, Bratislava.*

Birtles, A. B. and Wilkinson W. B. (In press). Mathematical simulation of groundwater abstraction from confined aquifers for river regulation.

Brereton, N. R. and Downing, R. A. 1975. Some applications of thermal infra-red linescan in water resource studies. *Wat. Services, 79,* 91-96.

Hamblin, M. J., Fisher, R. and Cluckie, I. D. 1975. Multi-site data generation for large water resource systems. *Int. Symp. on Application of Mathematical Models in Hydrology and Water Resources Systems, Bratislava.*

Nutbrown, D. A. 1974. Study of saline intrusion into Chalk on the south coast of the UK. Salt-Water Intrusion Meeting, Ghent. *Belgian National Committee for IHD.*

Nutbrown, D. A. (In press). Identification of parameters in a linear equation of groundwater flow.

Nutbrown, D. A. (In press). Normal mode analysis of the linear equation of groundwater flow.

Nutbrown, D. A., Downing, R. A. and Monkhouse, R. A. (In press). The use of a digital model in the management of the Chalk aquifer in the South Downs, England.

Reeves, M. J. 1975. A procedure for the prediction of nitrate levels in water supplies in the United Kingdom. *Int. Assoc. Wat. Poll. Res. Conference. Nitrogen as a water pollutant. Copenhagen.*

Reeves, M. J. 1975. A preliminary study of the relationship between cost and well field design in the Great Ouse Chalk groundwater development. *CWPU Tech. Note No. 4,* 27pp.

Reeves, M. J. 1975. Proposals for the development of the Triassic sandstone aquifer in the Calder valley, Cumbria. *CWPU Tech. Note, No. 5,* 15pp.

Reeves, M. J., Skinner, A. C. and Wilkinson, W. B. 1975. The relevance of aquifer flow mechanisms to exploration and development of groundwater resources. *J. Hydrol, 25,* 1-21.

Spears, D. A. and Reeves, M. J. (In press). The infiltration rate into an aquifer determined from the dissolution of carbonate. *Geol. Mag.*

Spears, D. A. and Reeves, M. J. (In press). The influence of superficial deposits on groundwater quality in the Vale of York. *Q. J. Engng. Geol.*

Van Oosterom, H., Downing, R. A. and Law, F. M. 1974. Development of the Chalk aquifer in the Great Ouse Basin. *Proc. Instn. Civ. Eng., 57,* 187-190.

Wright, C. E. 1974. The influence of catchment characteristics upon low flows in South East England. *Wat. Services, 78,* 227-230.

Wright, C. E. (In press). The assessment of regional groundwater schemes by river flow regression equations. *J. Hydrol.*

WATER DATA UNIT

Reading Bridge House, Reading RG1 8PS

This Unit was based on a nucleus of the former Water Resources Board's computer unit but strengthened by the addition of statisticians, chemists and biologists. It is the principal channel of information between the water industry and central government and deals with surface water, groundwater, water quality, financial and other statistics.
Its principal functions are:

a. To advise on the information required for water management purposes and the systems to be adopted for the collection, processing, storage and publication of such data;

b. To ensure the establishment of common standards and methods by the Regional Water Authorities; and

c. To collate and publish such data as is required on a national basis by Government, the National Water Council and the many interests affected by the activities of the water industry.

The Unit is currently supporting a number of research projects, some in association with other bodies; it is also the customer (in some cases a joint customer) in Rothschild terms, for research being undertaken at the Hydraulics Research Station and at certain NERC institutes.

Examples of the main publications are Water Data, the Groundwater Year Book, the Surface Water Year Book and the River Pollution Survey.

REGIONAL WATER AUTHORITIES

The hydrological research projects listed below, in alphabetical order of Authority, includes those projects originally initiated by the River Authorities during the period covered by this review.
The length of each project is indicated by an approximate starting date and, when known or estimated, the date of completion. A bibliography is provided of publications, or papers accepted for publication, arising from the projects.

ANGLIAN WATER AUTHORITY

Diploma House, Grammar School Walk, Huntingdon

Water balance in the Cam catchment
Great Ouse Rivers Division, AWA/Institute of Hydrology
1963-

The catchment of the River Cam is 200 km² in extent and is made up of highly pervious chalk strata, partly overlain by boulder clay.

The items in the water budget have been quantified by networks of raingauges, observation wells, soil moisture tubes, a river flow gauging weir and a climatological station.

The study shows that if a storage coefficient of 1·45% is adopted, the budget can be made to balance with reasonable accuracy.

Dickinson, W. T. and Douglas, J. R. 1972. A concepted runoff model for the Cam catchment. *Rep. No. 17, Inst. of Hydrol.*, 45pp.

Direct measurement of evapotranspiration
Essex River Division 1973-

Automatically controlled potential-evapotranspiration tanks have been installed alongside a conventional evaporation station. The P.E.T. figures obtained from these tanks will be compared directly with those obtained from the conventional station. It is intended that the experiments will be extended to include measurement of actual evapotranspiration from crops of varying root constants.

Prevention and effect of algal growth on river gauging structures
Essex River Division 1971

The first phase consisted of tests to determine whether or not a coating could be applied to gauging structures which would either reduce or inhibit algal growth.

The second was an investigation into the quantitative effect on flow measurement due to algal growth on river gauging structures.

Performance of crest tapping pressures on E.R.D standard weirs
Essex River Division 1972

The E.R.D. standard weir is a modification of the normal 'Crump' weir and so depends upon recording of the crest tapping pressures for assessment of non-modular flows. However, on occasions, these crest tapping pressures do not behave as expected, causing difficulties in calculations of the non-modular flows. Field tests are being carried out to investigate this phenomenon.

Hydrological operation of the Ely Ouse-Essex interbasin transfer scheme
Essex River Division 1970-

The Ely Ouse-Essex Scheme has now been operating for nearly three years. Over a substantial part of this period drought conditions of similar severity to the design conditions have prevailed. Work has been carried out to determine optimum hydrological operating rules; losses of transferred water through the bed of the river, flow and response times of transferred water (this has included the use of salt dilution techniques) and other practical operating data.

Huntingdon, R. and Armstrong, R. B. 1974. Operation of the Ely Ouse-Essex Scheme. *Ann. Gen. Mtg., Instn. civ. Engrs.*

NORTH-WEST WATER AUTHORITY

Dawson House, Great Sankey, Warrington, Lancs WA2 8JD

Operation of multi-source water supply systems, especially Lancashire conjunctive use scheme

P. D. Walsh 1970-1974

(as External Research Student, Department of Civil Engineering, Birmingham University)

The investigation has used critical period analysis and simulation modelling to determine and examine operating policies for a water supply system incorporating upland storage reservoirs, a lowland river intake in an unreservoired catchment, an inter-basin river transfer and groundwater abstraction from a Bunter Sandstone aquifer.

Walsh, P. D. 1971. Designing control rules for the conjunctive use of impounding reservoirs. *J. Instn. Wat. Engrs, 25* (7), 371-380.

Walsh, P. D. 1973. Investigation into the yield of an existing surface reservoir and aquifer system. Int. Fedn. for Automatic Control Symp. on Control of water resources systems, Haifa.

Walsh, P. D. 1973. The application of a simulation model to the planning and management of water resources in Lancashire. Wat. Res. Assn. Conf. on Computer uses in water systems. Contributed papers vol. 56-59.

Arley Brook Experimental Catchment: Estimation of deep percolation by water balance techniques

Rivers Division

Mersey and Weaver Unit 1973-1984

Systematic measurements are made of rainfall, runoff, groundwater levels in drift deposits and soil moisture variations in an impervious catchment. Actual evaporation is then inferred and its relationship with potential evaporation determined for use in neighbouring catchments with sandstone aquifers requiring conservation. The water balance for these areas should reveal a best assessment of deep percolation, a vital factor in resources management.

Collinson, P. A. J. 1973. The Arley Brook Experimental Catchment. Association of River Authorities Year Book and Directory 1973. 157-161.

Co-operative development of Bestel Dean interrogation and alarm raingauges

Rivers Division 1974-

Cumberland Unit

Co-operative project with manufacturers on development of instrument and warning system. The Bestel Dean instrument operates on a varying decay principle thus simulating soil moisture deficiency which is necessary for flood warning schemes.

Ground level raingauges

Rivers Division 1967-

Cumberland Unit

Differences in catch between standard and ground level gauges are recorded and evaluated with corresponding variation from accepted rainfall/runoff values.

Coalburn catchment study
Rivers Division 1966-1996
Cumberland Unit

An assessment of the effect on runoff from a catchment during the complete afforestation programme. Runoff from the natural catchment was observed until 1972 when the area was ploughed and drained prior to planting in the spring of 1973. The Institute of Hydrology is to assist until the end of 1977 when the study will revert to the Forestry Commission and NWWA.

Artificial freshet releases
Rivers Division 1971-75
Lancashire Unit

A controlled programme of freshet releases monitored by gauging weirs and fish counters is in hand to determine the optimum release flows and environmental conditions for stimulating salmon migration.

Nitrate consumption in river systems
Rivers Division 1975-
Lancashire Unit

Samples from the River Ribble and the River Lostock are being analysed to indicate behaviour of nitrate-polluted river water.

Paper mill effluent
Rivers Division 1973-76
Lancashire Unit

Analysis of samples are being used to obtain a better understanding of behaviour of paper mill effluent discharged to a river.

Investigation of the Sub-drift surface in Lancashire and Cheshire
Rivers Division 1972-
Mersey & Weaver Unit

Borehole data are used to delineate shape and depth of sub-drift depressions.

The age of waters below landfill sites
Rivers Division 1974-
Mersey & Weaver Unit

Tritium dating of groundwater from boreholes near tips to confirm recharge rates and flow net systems in the Permo-Triassic Sandstone aquifer.

Hydrogeology of Manchester area
Rivers Division 1971-74
Mersey & Weaver Unit

Data on the performance of the Permo-Triassic Sandstone aquifer in terms of yield and quantity are used for better management of the aquifer in Manchester.

Some aspects of the geochemistry of the Wirral Permo-Triassic Sandstone aquifer
Rivers Division 1972-75
Mersey & Weaver Unit

Groundwater quality data are collected for the construction of models to simulate saline water movement.

Bioassay for algal growth limiting nutrients
Eastern Division 1974-75
Manchester Water Supply Unit

Investigation of the relationship between biologically available nutrients determined by bioassay and nutrients determined chemically for better reservoir management.

NORTHUMBRIAN WATER AUTHORITY

Monitor House, Coast Road, Wallsend, Newcastle-upon-Tyne NE28 7RD

Development of a real-time operating system for the Kielder scheme

C. Drew 1973-1977
(Directorate of Planning and Scientific Services)

The development of a real-time, on-line control system which will incorporate the following features:

a. A simulation model of the hydrology of the catchments of the Rivers Tyne, Wear and Tees which will be used to predict river flows;

b. A system model and optimisation routine for optimal operation of the resource system;

c. A computer-controlled telemetry scheme to provide data for the system and simulation models and for automatic data reduction, archiving and retrieval.

Optimisation studies of water resource systems applied to the Kielder scheme

J. D. Hargreave (Sunderland Polytechnic, Control Engineering Dept.) and C. Drew (Directorate of Planning and Scientific Services, NWA) 1974-75

This project is initially concerned with the optimisation of the overall economic operation of the proposed Kielder scheme. The model is primarily a discrete form of the stochastic system, from which the optimal controls will be derived to minimise operating and treatment costs in conjunction with the maximisation of the stored resource. The study will be extended to consider the stochastic inputs to the system.

SEVERN-TRENT WATER AUTHORITY

Abelson House, Coventry Road, Sheldon, Birmingham B26 3PR

Storm station design

Avon Water Resources Division 1972-76

The objectives are to derive design criteria for storm stations constructed on trunk sewers from the analysis of data obtained from meteorological and hydrological instruments in the Coventry area.

Surface water study

Severnside Water Resources Division 1973-

A study of the effects of storms over an urban catchment (joint project with Gloucester C.C. and Bristol University).

Shropshire groundwater investigation

Severn River Management Division 1971-77

An investigation of the resources of the Shropshire Trias and its utilisation for river regulation.

Slope discharge method for flow gauging

Severn River Management Division 1974-

The project is designed to study the effects of weed growth on river flow.

Ultrasonic method of flow measurement
Severn River Management Division 1974-

An ultrasonic flow recorder has been installed at Ashleworth on the tidal River Severn experiencing wide variations in river level and sediment load.

Aeration studies
Upper Tame Water Reclamation Division 1974-

An investigation of the treatability of sewages containing metallic salts and industrial wastes and a study of the efficiency of nitrification at different temperatures.

Hydrogeology of the Bromsgrove area
Severn River Management Division 1974-

An assessment of the resources available in the Bunter Sandstone area.

Craig Goch scheme
Severn River Management Division 1973-

Experimental regulation releases and investigation of the ecological and chemical effects of the scheme.

Dover Beck scheme
Trent River Management Division 1974-

A study of the Nottinghamshire sandstone involving evaluation of ideas of conjunctive use and its effect on surrounding watercourses.

Flood forecasting
Trent River Management Division 1974-75

The development of a mathematical model to be linked to a telemetry network to predict flooding.

Sediment studies
Trent River Management Division 1974-

A study of sediment build-up on reservoirs and sediment transfer in experimental catchments. The project includes the development of a sediment sampler for tidal reaches in conjunction with Loughborough University.

Agricultural runoff
Trent River Management Division 1974-

Studies of nutrient concentrations from several monitored catchments.

Motorway runoff
Trent River Management Division 1974-

Analysis of runoff from motorways to monitor lead and phenolic concentrations.

Avoidance tests
Trent River Management Division 1971-

The development of a technique for measuring the impact of an effluent on a watercourse with regard to fishing.

Water quality
Trent River Management Division 1974-

Development of water quality monitors and telemetry network.

SOUTHERN WATER AUTHORITY

Guildbourne House, Chatsworth Road, Worthing, Sussex BN11 1LD

South Downs groundwater project
S.W.A. Resource Planning;
Water Research Centre;
Central Water Planning Unit;
Sussex River and Water Division;
Eastbourne Waterworks Company. 1970-80

Development of the chalk aquifer of the South Downs by the provision of additional point sources and extension of the controlled pumping system, thereby minimising the danger of increased saline intrusion from the coast and tidal waters. Programme includes extensive geophysical logging of deep coastal observation boreholes and a digital model study of the effects of different patterns and rates of pumping.

Water quality studies are in progress at the eastern end of the system where controlled dosing of nitrogen fertilizers is hoped to elucidate the effects of the latter on the nitrate content of Chalk groundwater.

South Downs Groundwater Project
(Brighton and Worthing Chalk Blocks)
First Progress Report November 1972.
Sussex River Authority/Water Resources Board

South Downs Groundwater Project
(Brighton and Worthing Chalk Blocks)
Second Progress Report 1975

Greene, L. A. and Walker, P. 1970. Nitrate pollution of Chalk waters. *Wat. Treatm. & Examn., 19,* 169-182.

Hardham Recharge Investigation
S.W.A. Resource Planning;
North west Sussex Water Division;
Water Research Centre;
Central Water Planning Unit. 1966-80

The investigation aims to determine the feasibility of recharging with river water a natural groundwater basin in the Folkestone Sands near Pulborough, Sussex, with a view to evaluating the combined river and groundwater storage resource. The adopted methods involve determination of the basin's hydrogeology, its relation to surface hydrology and the existing water quality; two series of recharge tests with raw and treated water through lagoons constructed in the aquifer; and trials of recharge via boreholes.

The Hardham Resource. 1971. Rept. Sussex River Authority/North-west Sussex Water Board.

Hardham Recharge Investigation:
First Progress Report 1971.
Second Progress Report 1972.
Third Progress Report 1973.
Sussex River Authority/Water Resources Board/North-west Sussex Water Board.

R. Itchen groundwater regulation scheme
S.W.A. Resource Planning 1973-80

The object is to study the feasibility and effects of regulating the flow of the Itchen by pumping groundwater from the Chalk into the headwaters of the three main branches, starting with the Candover stream. Particular stress is being laid upon the biological effects and on mathematical modelling to ensure the optimum use of the groundwater resources of the area. The Water Research Centre is making an independent study of the Itchen estuary which is hoped to define the minimum acceptable freshwater flow from which the extent of regulation needed can be calculated.

Itchen Groundwater Regulation Scheme
First Progress Report on Candover Pilot Scheme
Southern Water Authority, September 1974.

Tilmanstone Mine drainage study 1973-78
S.W.A. Resource Planning;
Water Research Centre;
Central Water Planning Unit.

A large quantity of saline water has been pumped out of Tilmanstone coal mine into the Chalk aquifer in the past. The project is a hydrogeological and mathematical modelling study of the feasibility and economics of rehabilitating the aquifer by removal of the contaminated water by controlled pumping. Particular interest is centred on confirming the apparent confinement of much of the contamination at a particular level, and exploiting this to remove the contamination in the most economical way.

SOUTH-WEST WATER AUTHORITY

Directorate of Resource Planning
Matford Lane, Exeter EX2 4QX

Development of automatic weather station and commercially available equipment
S. L. Paterson 1972-75

To develop the use of Normalair-Garrett automatic weather stations including solar radiation integrators.

Monthly rainfall-runoff conceptual modelling
S. I. Herbert, S. J. Booth 1974-75

To produce optimising models of the multiple store type for the purpose of generating long series of monthly river flows from rainfall data for Dartmoor and Exmoor rivers.

Time-series models of monthly runoff
T. S. Woodcock 1973-75

To develop monthly autoregressive time-series models for the Rivers Fowey and Tavy based on short historic runoff records. Check by use of rainfall-runoff multiple regression models.

Investigations into flooding of 10/7/68
S. L. Paterson, P. Kelway, 1969-71
S. I. Herbert

To analyse records of the storm to determine hourly rainfalls for a large number of grid points using A.R.I.D. techniques.

Sampling interval investigation for stage/flow measurement
B. D. Hooper 1975-

(i) To examine the frequency with which stage/flow measurements should be recorded at gauging stations.
(ii) To compare punched paper tape and autographic recorders as primary sources of data.
(iii) To filter the data before computer storage in order to retain the minimum acceptable quantity for future use.

Surface geophysical investigation of the Triassic aquifer of East Devon
SWWA (C. D. N. Tubb)/Department of Geological Sciences, University of Birmingham 1975

To investigate the topography and hydraulic parameters of the aquifer by a combination of geophysical techniques.

Hydrogeology and geochemistry of the Triassic aquifer of East Devon
SWWA (C. D. N. Tubbs)/IGS 1975-

(i) To explain the chemical variation within the aquifer in terms of groundwater movement and modification. (ii) To assist in the identification and prediction of suitable areas for further resource development.

Hydrogeology of the Permian aquifer in mid-Devon
Bristol University assisted by SWWA 1975-

To investigate the hydrogeology and potential for resource development of the Permian aquifer.

Hydrogeology of St Mary's, Isles of Scilly, and development of groundwater resources
SWWA (C. D. N. Tubb)/Department of Geological Sciences, Birmingham University (Dr Lloyd) 1975

To study the hydrogeology of the granitic island of St Mary's and prepare plans for extending the limited groundwater supplies.

Development of an automatic package for monitoring small scale pump tests
C. D. N. Tubb 1975

To provide a simple, compact package to monitor discharge and water level in boreholes during small scale pumping tests in order to improve supervision and data acquisition.

THAMES WATER AUTHORITY

New River Head, Rosebery Avenue, London EC1R 4TP

The definition of operating rules for the conjunctive use of resources
Dr D. J. Cook 1973-

The development of a suite of computer programmes to analyse the influence of various reservoirs, existing and planned, in the demand for supply in the Thames Water Authority area and to define rules for the operation of the system at a given risk of failure. The analysis includes a statistical investigation of the Thames flow record to predict the flow, for a given period, at the given risk of failure.

Proposed operational rule curves for conjunctive use of Thames catchment resources. 1975. *Thames Conservancy Int. Rep.*

The conjunctive use of Thames catchment resources: a definition of operational rule curves. 1975. *Thames Conservancy Division (TWA) Int. Rep.*

Estimation of recharge to the aquifer units of the Thames catchment
Dr D. J. Cook and B. J. D. Harris 1970-

The estimation of recharge, initiated as part of the Lambourn valley studies, has now been applied to all the aquifers of the catchment and the data is used in groundwater investigations and in determining catchment yields. A daily soil moisture analysis is carried out based on rainfall input and Penman estimates of potential evaporation. Output, in separated root constant zones and as an overall average, is actual evaporation and the percolation or recharge.

Electrical analogue model for groundwater studies
C. A. Hanson, J. R. Walker and
V. K. Robinson 1966-

The model is built to simulate the interaction between groundwater in an aquifer and as a component of stream flow. The model is single-shot, time dependent and slow time (100 days/sec) and the rivers exhibit bourne characteristics with the facility to reproduce stream bed leakage. The model, which is specific to a particular area, is excited by the input of current representing natural recharge and by the withdrawal of current by well simulators. Both groundwater level and stream flow can be monitored.

Mathematical model for groundwater studies
B. J. Connorton 1972-

Digital simulation of the hydraulic behaviour of an aquifer with particular reference to the Thames Ground Water Scheme is being developed using the method of successive over-relaxation. The model will be used in parallel with the electrical analogue model for a particular area and will be based on the same natural physical parameters. The model can be applied to any area by the appropriate representation of boundary conditions.

Regional variation in aquifer hydraulic properties and their geological prediction
M. Owen, S. Bennett and B. J. Connorton 1973-76

Transmissivity and storativity have been determined by standard analytical techniques from pumping tests in the area of the first stage of the Thames Ground Water Scheme and regional interpretations of the parameters have been made. These interpretations will be compared with relatively subjective interpretations based on simple, empirical hydrogeological criteria with very limited control from field data. Having made any necessary modifications to the subjective interpretation, it will be applied to areas for further development of the scheme in order to provide good predictions of transmissivity and storativity to build into extensions of the analogue and digital aquifer models.

Assessment of borehole performance
B. J. Connorton, S. Bennett and W. Price, TWA.
K. Rushton, Birmingham Univ. (Civil Eng.) 1973-76

The development of drawdown in a pumped well in a fissured aquifer frequently does not conform to analytical models derived for idealised aquifers. Using hydraulic and geophysical data from pumping tests of approximately 20 days' duration, techniques are being developed to extrapolate this test data to an operational condition of approximately 250 days, in terms of variation in both pumping rate and drawdown. The techniques being investigated are modifications to long-established methods and to a method recently developed by Dr Rushton, to allow for the characteristics of the flow mechanism and hydraulic properties of the aquifer.

Regional geophysical investigations of inflow horizons in Chalk and Jurassic limestone aquifers
V. K. Robinson and T. N. Woods 1972-

In the Chalk and Jurassic limestone aquifers water enters boreholes at a limited number of discrete horizons related to fissures. Using temperature, conductivity and flow meter logging, the distribution of these inflow horizons in relation to depth below surface, depth below water table, lithology, stratigraphy and electrical resistivity marker bands has been investigated. In the Chalk of the Berkshire Downs, where intensive drilling and test pumping of boreholes has been carried out as the first part of the Thames Groundwater Scheme, the results have been used in combination with other data to postulate an overall flow mechanism for the aquifer.

Induced recharge of river water into the Chalk at Taplow, Berks.
M. Owen, A. E. Jones and V. K. Robinson, TWA
W. M. Edmunds, T. K. Tate and A. Robertson, IGS
1968-74

Groundwater chemistry, temperature and conductivity logging of the borehole water column and hydraulic analysis of pumping tests by standard methods were used to quantify the proportion of induced recharge of river water in the discharge from pumping boreholes.

W. M. Edmunds, M. Owen and T. K. Tate. (In press). Estimation of induced recharge of river water into Chalk boreholes at Taplow, Berkshire, using hydraulic analysis, geophysical logging and geochemical methods. *Inst. geol. Sci., Res. Rep. Series.*

Hydrogeological maps of the Chalk aquifer of the western London basin and the Jurassic limestone aquifers of the Cotswolds
M. Owen, B. J. Connorton, M. D. Eggboro
and V. K. Robinson 1971-

Interpretations of hydrogeological data from current and historical records and from field surveys are presented in map form in terms of spatial distribution of aquifers, positions of water tables and pressure surfaces and any long term changes to them, groundwater chemistry parameters and hydrogeological classification of streams and springs.

Development of inexpensive water well logging techniques
V. K. Robinson 1970-73

A water well geophysical logging system based on equipment originally developed by I.G.S. has been extended to provide the capability for continuous running of electrical logs previously carried out in a discontinuous mode.

Robinson, V. K. 1974. Low cost geophysical well logs for hydrogeological investigations. *Quart. J. eng. Geol.*, *7*, 207-15.

Time series: Generation of synthetic discharges
R. J. Mander, S. C. Childs, A. E. Jones
and Mrs D. Richards 1971-

The objective is the synthesis of hydrographs of total

discharge, or of groundwater discharge, to provide long periods of daily data from limited periods of observation, for water resources requirements generally. The method differs from those in use up to the present time in being a physical rather than conventional stochastic interpretation, and is based upon the theory that the outflow from a single natural system of storage, or from a combination of such systems, will behave always in accordance with a basic physical law, the characteristics of which are unique to any given natural system, single or combined, as a statistical norm. Under conditions of zero disturbance the outflow characteristics are apparent as a normal recession but, under natural conditions of flow, disturbances are continuously causing the discharge to deviate from the normal recession. These disturbances, as 'input equivalents', can be abstracted from an historic record and their independent probability distribution determined by consideration of initial energy states. Synthesis of values having similar properties is then possible, enabling the generation of discharge to be achieved.

Mander, R. J. 1971. Stochastics of Non-Linear Storage. *Thames Conservancy Int. Rept.*

Mander, R. J. 1974. Discharge from Compounded Storage. *Thames Conservancy Int. Rept.*

Conceptual models of groundwater storage and flow

R. J. Mander, B. J. Greenfield and
Miss C. Hyoms 1969-

Development of a general conceptual model relating point groundwater levels to each other and to groundwater discharge as a surface manifestation. Derivation of groundwater levels and groundwater component of streamflow from recharge data. The present model is currently applied both as a predictive tool and for water resources purposes generally. Development continues with special regard to the groundwater development proposals.

Mander, R. J. 1971. A Concept of Unconfined Aquifer Storage and Discharge. *Thames Conservancy Rept.*

Mander, R. J. and B. J. Greenfield. 1974. Unconfined Storage and Discharge. *Thames Conservancy Int. Rep.*

Conceptual models of catchment total runoff

R. J. Mander, F. A. K. Farquharson and
Mrs D. Richards 1969-

The two main objectives are to provide predictions of tributary catchment runoff from rainfall, primarily for flood forecasting purposes and also to describe simultaneously runoff as tributary catchment totals and cumulative runoff in terms of lock reaches along the Thames to Teddington.

Channel routing by conceptual model

R. J. Mander and B. J. Greenfield 1973-

The concept is applied to routing of flows in regulated channels. Current work relates to the Thames abstraction reach between Datchet and Teddington and is intended as an operational tool to aid in efficient control of abstractions for water supply. At the present stage the model covers four of the eight abstraction reaches, and is being developed to cover the remainder. It is planned to extend the model to reaches further upstream.

WELSH NATIONAL WATER DEVELOPMENT AUTHORITY

The Barracks, Brecon LD3 7HP

Catchment models based on ISO-function
A. O. Lambert, Dee and Clwyd River Authority 1969-

Development of mathematical models for predicting catchment response to rainfall, using the water balance equation and considering that catchment storage/outflow relationships are made up of both logarithmic and linear segments. Extension of this type of model to other Inflow-Storage-Outflow problems.

Lambert, A. O. 1972. Catchment models based on ISO-Functions. *J. Inst. Wat. Engrs, 26* (8), 413-422.

Lambert, A. O. 1969. A comprehensive rainfall/runoff model for an upland catchment area. *J. Inst. Wat. Engrs, 23* (4), 231-238.

Evaluation of open channel fish counting on a large river
D. Simpson, A. Clarke (North of Scotland Hydro Electric Board) and
R. I. Millichamp, Usk River Authority 1973-75

To determine the effectiveness of open channel fish counting on a large river using resistivity methods and the optimum arrangement and number of detecting strips and counters.

Millichamp, R. I. 1974. *J. Inst. Fish Mgmt, 5* (2), 55, 56.

Extension of river flow records for planning development
Dr M. J. D. Thoms and A. S. Wain
Usk River Division 1971-73

Flow records of the river Usk do not cover years with severe droughts; records for such years were necessary to evaluate the yield of the area's water resources. Records were synthesised for such years by the 'equiprobability' technique, which relates the flow frequency curves of two rivers – one having a long record – and uses the relationships derived to simulate a hydrograph for the design periods required for water resources evaluations.

Usk River Authority. 1972. 'Report on survey of water resources'.

Thoms, M. J. D. and Wain, A. S. (In press). 'Extension, and extended use of river flow records'. *J. Inst. Wat. Engrs.*

WESSEX WATER AUTHORITY

Techno House, Radcliffe Way, Bristol BS1 6NY

Gussage bed stream lining 1971-76

Various methods of sealing the bed of a bourne stream which, under natural conditions, goes dry each year, are under investigation. Puddled chalk, dry chalk, butyl and polythene sheeting have been used as sealants and the efficiency of each method is being monitored by measuring flows over weirs along the channel. A special drainage system has been incorporated in the bed to counteract the high uplift pressures in winter. The results will determine the suitability of regulating bourne streams artificially from groundwater abstractions.

Groundwater studies in the Cotswold and Mendip hills 1970-76

Groundwater levels are regularly monitored and water tracing is used to establish underground catchments and sources of pollution. The effects of limestone quarrying in the Mendips on groundwater sources are being studied to enable predictions to be made should further quarrying be contemplated. In the Great Oolite aquifer an observation borehole network has been established and prolonged abstraction pumping tests are proposed. With associated river regulation following the testing for leakage in river beds the Cotswolds could provide a significant source of water for public supply.

Automatic climatological station 1968-

An assessment of the performance of an automatic climatological station at Chew Stoke compared with the results achieved by conventional means. The application of this type of station is by the production of Penman potential evapotranspiration figures which may be applied to the catchment, particularly to assess percolation and yield.

Tilting weirs as permananent gauging structures 1974-76

There is extensive use of tilting weirs in lowland Somerset to maintain river levels but no accepted method of obtaining records of flows over these weirs. Instrumentation is being developed to obtain continuous records of river flows over these land drainage structures. Water level recorders are used to measure the angle of tilt, crest level, upstream and downstream water levels and the structures are rated by model and by current meter.

YORKSHIRE WATER AUTHORITY

21 Park Square South, Leeds LS1 2QG

Investigation into hydrogeology of Triassic Sandstone in the Vale of York

Yorkshire River Authority;
Water Resources Board 1970-1974

To ascertain the yield of the Sandstone aquifer with particular reference to river regulation.

Preliminary investigation into the geology and hydrogeology of the Vale of Pickering

Yorkshire River Authority;
Water Resources Board 1972-1974

Sinking of exploration and observation boreholes and construction of streamflow gauges to ascertain the yield of the Corallian aquifer with particular reference to river regulation.

Trunks of trees are cut off under water during investigations into the biological controls of evaporation.

Institute of Hydrology

CHAPTER IV

HYDROLOGICAL RESEARCH IN UNIVERSITIES IN THE UNITED KINGDOM

Distribution of Hydrological Studies

University interest in hydrology is widespread but no specialist department of hydrology has yet been created. Hydrological research is carried out by individuals working in a variety of departments where hydrology is taught at undergraduate or postgraduate levels. Most first degree courses in civil engineering, environmental science or geography and many courses in agricultural science, biology and geology include some hydrology. At the postgraduate level, advanced courses in hydrological subjects are offered at several universities as listed in Table 6.

Layout

University research projects are listed below in alphabetical order of universities. Beside the name of each university department an abbreviation is given for ease of cross-reference from the tabular statement in Chapter II where the projects are classified according to subject.

A brief description of each project is given with the names of the research workers concerned. The names of the research students and research assistants are given in brackets following the names of the project supervisors. The length of each project is indicated by an approximate starting date and, when known or estimated, the date of completion. A bibliography is provided of publications, or papers accepted for publication, arising from the projects since 1970.

UNIVERSITIES

ABERDEEN

Department of Botany (ABDBT)

Carbon dioxide and water vapour exchange in a forest canopy

Dr P. G. Jarvis, G. B. James, Dr R. E. Neilson, Dr W. R. Watts (B. Davidson, G. P. Richards) 1968-75

Replicated gradients of water vapour, carbon dioxide concentration and air temperature are measured above the canopy together with net radiation and air and plant temperatures within the canopy to give fluxes of sensible heat, latent heat and carbon dioxide. The fluxes are partitioned within the canopy by diffusion porometry and $^{14}CO_2$ uptake by the foliage.

Beardsall, M. F., Jarvis, P. G., Davidson, B. 1972. A null balance diffusion porometer suitable for use with leaves of many shapes. *J. appl. Ecol., 9,* 677-690.

Hellkvist, J., Richards, G. P., Jarvis, P. G. (In press). Vertical gradients of water potential and tissue water relations in Sitka spruce trees, measured with a pressure chamber. *J. appl. Ecol., 11.*

Jarvis, P. G., James, G. B., Landsberg, J. J. (In press). Coniferous forest. In: *Vegetation and the atmosphere* (Ed: J. L. Monteith). Academic Press, London and New York.

Landsberg, J. J. and Thom, A. S. 1971. Aerodynamic properties of a plant of complex structure. *Quart. J. roy. met. Soc., 87,* 565-570.

Landsberg, J. J. and James, G. B. 1971. Wind profiles in plant canopies: studies on an analytical model. *J. appl. Ecol., 8,* 729-741.

Landsberg, J. J., Jarvis, P. G., Slater, M. B. 1972. The radiation regime of a spruce forest. In: *Plant response to climatic factors* (Ed: R. C. Slatyer). *Ecology & Conservation, 5,* 411-418, UNESCO, Paris.

Landsberg, J. J., Jarvis, P. G. 1973. A numerical investigation of the momentum balance of a spruce forest. *J. appl. Ecol., 10,* 645-655.

Norman, J. H., Jarvis, P. G. (In press). Photosynthesis in Sitka spruce (*Picea sitchensis* (Bong.) Carr.) III. Measurements of canopy structure and interception of radiation. *J. appl. Ecol., 11.*

Photosynthesis and transpiration in Sitka spruce in relation to genotype and environment

Dr P. G. Jarvis, Dr R. W. Neilson with Dr N. C. Turner (1973) (C. L. Beadle, Miss M. Lewandowska) 1960-

Exchanges of water vapour, carbon dioxide and energy at the leaf-air interface of Sitka spruce seedlings are being investigated in the laboratory to give physiological information for models of forest processes.

Cornic, G., Jarvis, P. G. 1972. Effects of oxygen on CO_2 exchange and stomatal resistance in Sitka spruce and maize at low irradiances. *Photosynthetica, 6,* 225-239.

Landsberg, J. J., Ludlow, M. M. 1970. A technique for determining resistance to mass transfer through the boundary layers of plants with complex structures. *J. appl. Ecol.*, *7*, 187-192.

Ludlow, M. M., Jarvis, P. G. 1971. Photosynthesis in Sitka spruce (*Picea sitchensis* (Bong.) Carr.) I. General characteristics. *J. appl. Ecol.*, *8*, 925-953.

Jeffree, C., Johnson, R. P. C., Jarvis, P. G. 1971. Epicuticular wax in the stomatal antechamber of Sitka spruce and its effects on the diffusion of water vapour and carbon dioxide. *Planta* (Berl.), *98*, 1-10.

Department of Geography (ABDGG)

Hydraulic geometry, sediment transport, coastal morphology and pollution dispersion studies in the Dee, Don and Ythan estuaries

Prof. K. Walton, Dr W. Ritchie
(G. C. Stove, G. H. Buchan, M. Rose) 1972-76

Hydraulic aspects of pollution dispersion are being studied in the Don and Ythan estuaries. Water flow and sediment transport are being studied in Aberdeen harbours before and after engineering works and attention is being given to the changing nature of a spit-bar shoreline at the mouth of the Don. New techniques of X-ray diffractometric analysis of suspended sediments and heavy mineral separation are being used in the Ythan estuary to relate sediment movement to hydraulic geometry. Freshwater budgets are included in these studies.

Department of Geology and Mineralogy (ABDGL)

Hydrogeology of Knock Hill, Banffshire

G. Downie (G. G. Brebner) 1973-74

Continuous airlift pumping from 10m below the static water level in bores was used to locate a possible well with a $4m^3/hr$ flow.

Preliminary investigation of springs on Ben Rinnes

G. Downie 1973-74

Flow rates from springs issuing from jointed and weathered granite are being measured.

Geophysical investigation of the Alford Basin, Aberdeenshire

Dr W. A. Ashcroft, G. Downie,
J. T. C. Hay (D. Sirikci) 1974

The extent and thickness of unconsolidated sediment in Pleistocene deposits will be measured by seismic refraction and resistivity techniques to estimate the value of the Alford Basin as a water reservoir.

Department of Soil Science (ABDSS)

Water regimes in gley soils

Dr D. A. MacLeod (Dr T. R. Moore) 1969-72

The relationship between gley development and present-day water regime was studied on 52 soils.

Moore, T. R. 1973. The distribution of iron, manganese and aluminium in some soils from north-east Scotland. *J. Soil Sci.*, *24*, 162-71.

Effects of water supply and other site factors on growth of grass and Sitka spruce in north-east Scotland
Dr D. A. Macleod (Dr J. F. Blyth) 1970-74

Soil water regimes, with or without secondary drainage treatment
Dr T. Batey, Dr C. E. Mullins (H. L. A Htun) 1974-

A neutron probe and tensiometers are used to study regime of mole-drained, subsoiled and undrained plots on a Pitmedden series soil of the Taves association.

ASTON IN BIRMINGHAM

Department of Civil Engineering (ASTCE)

Mathematical models of precipitation, evaporation and snowmelt runoff in mountainous regions
W. J. Walley 1971-

The effect of relief on hydrology is modelled to improve evaluations of yield from upland catchments where only limited data are available.

Hutchinson, P., Walley, W. J. 1972. Calculation of areal rainfall using finite element techniques with altitudinal corrections. *Hydrol. Sci., Bull, 17*, 310.

Computer simulation of the hydrologic cycle
Dr T. R. E. Chidley, Dr R. Johnson
(E. Smith) 1971-74

An integrated simulation of surface and groundwater hydrology is used to predict flow and some techniques of parameter identification have been investigated.

BELFAST, QUEEN'S

Department of Geography (BQUGG)

River flood and water supply problems in Northern Ireland
Dr R. Common 1970-74

Physical characteristics and social surveys form the basis for water resource appraisal in a number of troublesome catchments.

BIRMINGHAM

Department of Civil Engineering (BRMCE)

Data generation
Prof. M. J. Hamlin 1968-75

Research is being continued on extension of short records to produce pseudo-historic records based on longer-term data derived from adjoining catchment areas and extended to the simultaneous generation of compatible synthetic data sets for up to ten sites where it is necessary to preserve the serial and cross correlation statistics.

Hamlin, M. J. 1971. A study of synthetic flow generation techniques using Elan Valley data. *J. Instn. wat. Engrs., 7*, 355-370.

Hamlin, M. J., Kottegoda, N. T. 1973. The preparation of a data set for hydrologic system analysis. *Int. Ass. scient. Hydrol. Symposium, Madrid, 1*, 163-177.

Determination of travel time in rivers
Prof. M. J. Hamlin (A. Tillotson) 1971-74

A need to determine travel times in rivers when the flow is below average arises more frequently as the number of regulation or transfer schemes increases. There are also cases where because of accidental spillage of toxic materials, it is important to know what are travel times in the lower reaches of a river. Work so far has been based on data from the Lancashire River Division of the North West Regional Water Authority but may be extended to other river systems in Britain.

A stochastic model for the prediction of dissolved oxygen in a polluted river
Prof. M. J. Hamlin (N. M. D. Green) 1970-73

The dynamic situation in a river is considered using a stochastic modelling procedure to predict the distribution of dissolved oxygen levels downstream of a pollution discharge. A simulation model has been developed and a two-month survey made of the River Avon, downstream of the discharge point of Finham sewage treatment works.

Runoff from urban catchments via surface channels and stormwater sewers
Prof. M. J. Hamlin (S. R. Wood, W. N. A. Hindi) 1973-

Surface runoff in urban areas and changes caused by increasing development have been modelled (S. R. Wood). Stormwater sewerage systems are being modelled with particular reference to the effect of moving storms, varying infiltration rates and the pattern of storms associated with different synoptic situations (W. N. A. Hindi).

Derivation of optimal control rules for complex water resource systems
Prof. M. J. Hamlin (I. Cluckie) 1971-75

The determination of optimal control rules for complex water resource systems is often solved by a series of computer simulations for a range of possible solutions. This can be costly and is often inefficient in terms of use of computer time. Control theory is complex and solutions which have been derived for electrical or production engineering are often not suitable for water resource systems. This research is primarily concerned with river regulation systems where the controlling reservoir has both a natural and a pumped inflow.

Hamlin, M. J., Kottegoda, N. T., Kitson, T. 1973. Control of a river system with two reservoirs. *IFAC Symposium on control of water resources systems*, Haifa, Israel, 1-20.

Investigation into rainfall characteristics at a point and over areas
Prof. M. J. Hamlin (Dr P. S. Kelway, Mrs J. C. Baugh) 1973-76

Available rainfall analysis techniques will be applied to a comprehensive study of a number of rainfall events. Rainfall distributions will be analysed over short-term intervals and mesoscale systems will be studied in detail; comparisons will be made of the rainfall patterns identified with those derived from radar analyses. Depth-area-duration information will be obtained for each event.

Kelway, P. S. 1974. A scheme for assessing the reliability of interpolated rainfall estimates. *J. Hydrol., 21,* 247-267.

Regional flood analysis based on stochastic correlation and physical parameters

Prof. M. J. Hamlin, Dr N. T. Kottegoda
(A. J. Girling) 1973-74

Flood data is studied above a threshold value from subcatchments in the Wye basin, and separately for other areas such as the Welland and Nene basins. Temporal and spatial correlations, distributions of the mean peak heights and number of peaks, mathematical fits to same, physiographic factors and climatic conditions are to be incorporated in a multivariate model for generation of flood data.

Kottegoda, N. T. 1972. Flood evaluation by some data generation techniques. *Proc. int. hydrol. Symp.* 591-594. *Colorado State University, Fort Collins, U.S.A.,* 189-199.

Kottegoda, N. T. 1972. Flood evaluation – can stochastic models provide an answer? *Proc. int. Symp. on Uncertainties in Hydrologic and Water Resource Systems, University of Arizona, Tucson,* 105-114.

Extension of flow records through stochastic models and application to hydrologic systems

Dr N. T. Kottegoda 1971-

Stochastic models are used for generation of pentad flow data in univariate, bivariate, and multivariate cases. Study is made of crossing properties, deficits, surplus flows, probability distributions and seasonal variations. The time series of synthetic data are used in hydrologic systems analysis, reservoir designs and control rules for conjunctive use. Daily data are being studied.

Kottegoda, N. T. 1972. Stochastic five daily streamflow model. *J. hyd. Div. Amer. Soc. civ. Engrs., 98.* (HY9), 1469-1485.

Hamlin, M. J., Kottegoda, N. T. 1973. The preparation of a data set for hydrologic systems analysis. *Proc. Symp. on the Design of Water Resource Systems with Inadequate Data, Madrid, 1973, UNESCO,* 163-177.

Hamlin, M. J., Kottegoda, N. T., Kitson, T. 1973. Control of a river system with two reservoirs. *IFAC Symp. on Control of Water Resources Systems, Haifa, Israel.*

Kottegoda, N. T. (In press). The effects of skewness in three stochastic pentad riverflow models on crossing properties of synthesized data. *Wat. resour. Res. Wash.*

Unconfined-confined flow close to wells and rivers

Dr K. R. Rushton (Mrs A. Turner, D. Elgy) 1972-

When the quantity of water pumped from wells causes significant drawdowns, the problem becomes nonlinear. A numerical analysis has shown within what ranges the commonly-used approximations are valid. Also a three-dimensional electrical analogue is being used to model the free surface behaviour of well-river systems, and the reliability of two-dimensional approximations to these problems is being examined.

Rushton, K. R. 1973. Extensive pumping from unconfined aquifers. *Wat. resour. Bull., 9.*

Rushton, K. R., Turner, A. (In press). Numerical analysis of pumping from confined-unconfined aquifers. *Wat. resour. Bull.*

Analysis of groundwater flow
Dr K. R. Rushton, Dr R. H. Herbert,
Mrs L. M. Tomlinson (Dr L. Wedderburn,
R. G. Bannister, I. Chan) 1966-

Analogue and digital computers are used both to model seepage flow around dams and other obstacles and to investigate regional groundwater flow. Steady and time variant problems have been considered. Certain of the widely-used digital computer methods have been shown to lead to erroneous results. Analogue models controlled by a small digital computer have proved valuable in representing periods of hundreds of years.

Rushton, K. R., Herbert, R. 1970. Resistance network for three-dimensional unconfined groundwater problems with examples of deep well dewatering. *Proc. Instn. civ. Engrs., 45,* 471-490.

Rushton, K. R., Bannister, R. G. 1970. Aquifer simulation on slow time resistance-capacitance networks. *Groundwater, 8,* 15-24.

Rushton, K. R., Tomlinson, L. M. 1970. Digital computer solutions of groundwater flow. *J. Hydrol., 12,* 339-362.

Rushton, K. R., Wedderburn, L. A. 1971. Aquifers changing between the confined and unconfined state. *Groundwater, 9,* 5.

Rushton, K. R., Wedderburn, L. A. 1973. Starting conditions for aquifer simulations. *Groundwater, 11,* 37-42.

Rushton, K. R. 1973. Discrete time steps in digital computer analysis of aquifers containing pumped wells. *J. Hydrol., 18,* 1-19.

Rushton, K. R. 1973. Critical analysis of the alternating direction implicit method of aquifer analysis. *J. Hydrol., 21,* 153-172.

Rushton, K. R. (In press). Aquifer analysis using backward difference methods. *J. Hydrol.*

Study of an unconfined aquifer behaviour
Dr T. D. Streltsova 1971-

The behaviour and delayed response of an unconfined flow to a well (axisymmetrical case) and to a reservoir (two-dimensional case in a cross-section) is considered on a basis of recognition of the variable vertical movement of flow taking into account the elastic properties of an unconfined aquifer, drainage from the unsaturated zone and the partial penetration of the abstraction well.

Streltsova, T. D. 1972. Unsteady radial well flow equation in an unconfined aquifer. *Proc. Instn. civ. Engrs., 51,* 362-362.

Streltsova, T. D. 1972. Unconfined aquifer and slow drainage. *J. Hydrol., 16,* 117-127.

Streltsova, T. D. 1972. Unsteady radial flow in an unconfined aquifer. *Wat. resour. res. Wash., 8, 4,* 1059-1066.

Streltsova, T. D., Rushton, K. R. 1973. The water-table drawdown due to a pumped well in an unconfined aquifer. *Wat. resour. Res., Wash.*, 236-242.

Streltsova, T. D. 1973. The flow near a pumped well in an unconfined aquifer under non-steady conditions. *Wat. resour. Res., Wash.*, 227-235.

Streltsova, T. D. 1973. On the leakage assumption applied to equations of groundwater flow. *J. Hydrol. 20, 3,* 237-253.

Aquifer test design for pumping in the vicinity of a river

Dr T. D. Streltsova (M. W. Kawecki) 1973-74

Behaviour of flow to a well near a river is studied for the rationalization of pumping test programmes by detailed investigation of the type and extent of the hydraulic interaction between the stream and the aquifer, the influence of the partial penetration of the river bed, the effect of semi-perviousness of the river bed.

Streltsova, T. D. 1972. The basis of the method of seepage resistance. Synopsis. *Proc. Instn. civ. Engrs.*, 597.

Streltsova, T. D. (In press). Method of additional seepage resistance – theory and application. *Proc. Am. Soc. civ. Engrs. Hydraul. Div.*

Application of analytic flood routing methods to British rivers

Dr I. E. Thomas (Dr P. R. Wormleaton, A. Milburn) 1968-74

Following earlier work on the diffusion model of flood movement in rivers, the effectiveness of applying hydraulic equations to unsteady flow to both uniform and natural channels has been assessed. Solutions by digital computer have been developed and quantitatively compared.

Thomas, I. E., Wormleaton, P. R. 1970. Flood routing using a convective diffusion model. *Civ. Engrg. publ. Works. Rev.*

Thomas, I. E., Wormleaton, P. R. 1971. Finite difference solution of the flood diffusion equation. *J. Hydrol., XII* (3), 211-220.

Department of Geography (BRMGG)

Soil moisture variability in the Forest of Dean

Prof. R. P. Moss (E. M. Bignall) 1971-74

The variability of soil moisture in relation to soil type, slope angle, slope position and vegetation is being examined in two valleys in Forestry Commission land, using regular sampling along transects with gravimetric determination of moisture, supplemented by data from gypsum blocks. Multivariate analysis of the data is being used as the basis for prediction of soil moisture values for slope, soil and vegetation parameters. These predictions are then to be tested against actual measurements.

Groundwater conditions in the Peak District

Dr D. P. Ede 1974-

Groundwater analysis and dye tracing will be used to trace flow patterns within the Carboniferous Limestone area.

Department of Geology (with Geophysics) (BRMGL)

Hydrogeology of the Keuper Sandstone in the Droitwich syncline area, Worcestershire
Prof. F. W. Shotton (T. Ramingwong) 1972

Groundwater resources assessment of the Keuper Sandstone aquifer.

Geophysical investigations in the Shropshire Trias relating to groundwater development
Prof. D. H. Griffiths (M. Bilbo) 1972-76

Gravimetric, seismic and electrical resistivity methods have been employed to elucidate the geological structure and the nature of glacial drift in the southern part of the Cheshire Basin. This work is in connection with a major groundwater development scheme for river regulation being carried out by the Severn-Trent Water Authority.

Investigation of some electrical properties of Permo-Triassic sandstones
Prof. D. H. Griffiths (F. A. Collar) 1972-

The induced polarization properties of sandstone samples have been investigated in the laboratory to produce a geophysical technique that will, by surface measurements, define the most permeable areas of a sandstone aquifer. A useful correlation has been found to occur in limited conditions between IP and permeability; field testing is being undertaken.

Geophysics applied to problems in hydrology
Prof. D. H. Griffiths (J. R. Forshaw) 1972-75

The use of both P and S seismic wave velocities for the determination of hydrological properties of sandstone is being investigated. Primary relationships have been observed in the laboratory between intergranular permeability and velocity, using ultrasonic pulse techniques. The work has been extended to field conditions in the Bunter Sandstone of the Fylde area of Lancashire.

Hydrogeological assessment of the central Lincolnshire Limestone
Dr J. W. Lloyd (D. W. Peach) 1973

A variety of techniques are used to assess groundwater resources and the effects of development on saline water present in the aquifer.

Radiocarbon dating of groundwater
Dr J. W. Lloyd 1974

Carbonate precipitates are sampled in the field and methane prepared in the laboratory for measurement by proportional gas counting to obtain groundwater age profiles in major British aquifers.

Trace element groundwater studies
Dr J. W. Lloyd, Dr G. Hendry 1974-

The use of trace elements in the definition of groundwater movement is being evaluated.

BIRMINGHAM

Department of Physics (BRMPH)

Physics of ice
Dr J. W. Glen (Dr J. G. Paren, G. C. Camplin, D. R. Homer, W. J. Fitzgerald) 1963-77

The effect of grain boundaries, preferred crystal orientation and impurities on the mechanical and electrical properties of ice is being studied to improve understanding of the mechanical response of ice to loading and to interpret electrical sounding of natural ice masses.

Camplin, G. C., Glen, J. W. 1973. The dielectric properties of HF-doped single crystals of ice. *Physics and chemistry of ice: Symp. on the Physics and Chemistry of Ice, Ottawa, Canada, 1972* (*Eds. E. Whalley, S. J. Jones, L. W. Gold*). Ottawa, Royal Society of Canada. 256-61.

Glen, J. W. 1970. Thoughts on a viscous model for sea ice. *AIDJEX Bulletin, 2,* 18-27.

Glen, J. W. 1972. Total melting time in the ablating slab problem. *J. appl. Phys., 43, 1,* 258.

Glen, J. W. 1974. *The physics of ice. Cold Regions Science and Engineering. Monograph* II – C2a. Hanover, New Hampshire, Corps of Engineers, U.S. Army. Cold Regions Research and Engineering Laboratory. *7.* 81.

Paren, J. G. 1973. The electrical behaviour of polar glaciers. *Physics and chemistry of ice: Symp. Phys. & Chem. Ice, Ottawa, Canada, 14-18 August 1972.* (*Eds. E. Whalley S. J. Jones, L. W. Gold.* Ottawa, Roy. Soc. Canada, 262-67.

BRADFORD

Postgraduate School of Studies in Civil Engineering (BRDCE)

Evaporation studies
M. F. Baty 1970

During-the-year and year-to-year variations of the coefficient relating evaporating pan evaporation to the Penman estimate of openwater evaporation are being studied together with the effect of the Penman estimate of modifications to the subsets in the set of constant parameters.

Baty, M. F. (In press). A comparative study of the correlation between differing techniques for obtaining evaporation data derived for several different areas of the world, when applied to arid climates. *Int. Sem. Expn. Wat. Res. Instrumn. 1974. Int. Wat. res. Assn. Chicago, 1974.* 27pp.

Hydrology of small catchments
M. F. Baty (M. R. Hasan) 1969-70

Two catchments of different permeabilities in the Lee Basin were compared.

BRIGHTON POLYTECHNIC

Department of Civil Engineering and Building (BRPCE)

Analysis of rainstorms over Sussex
Dr F. D. C. Henry (J. E. Dent) 1970-74

Records from extensive networks of autographic and daily raingauges are analysed in relation to synoptic conditions to determine the influence of the south coast and of the South Downs.

Mobilization and transport of sediment on slopes on Chalk
Dr F. D. C. Henry (S. C. Oakes) 1970-74

Permanent troughs have been installed at 6 sites to collect surface and shallow interflow to correlate loads with rainfall, and other meteorological and geomorphological data.

Sediment transport in streams in Sussex
Dr F. D. C. Henry (Miss P. Zalasinski) 1970-74

Permanent suspended, dissolved and bed load samplers have been installed in streams to correlate load measurements with hydrological regimes.

BRISTOL

Department of Civil Engineering (BTLCE)

Dispersion induced by density effects and sedimentation
Dr T. L. Shaw (S. Buttling) 1964-76

Experimental and theoretical studies of dispersion induced by thermal, solution and suspension effects are being carried out. Particular attention is being given to the influence of boundary geometry on patterns of flow, including sedimentation.

Buttling, S., Shaw, T. L. 1973. Predicting the rate and pattern of storage loss in reservoirs. *11th Congress Int. Ass. on Large Dams*, 569-584.

Shaw, T. L., Buttling, S. 1971. Sedimentary processes in reservoirs. *14th Cong. Int. Ass. hydraul. Res., 5,* Paper 109, 31-34.

Attenuation of storm runoff by flood plains
Dr T. L. Shaw (S. W. Huntington) 1969-74

Flood wave attenuation in streams has been considered, experimentally and analytically. Flood plains have also been studied, to assess optimum shape and position of flood areas relative to urban areas, tributaries.

Shaw, T. L., Huntington, S. W. Flood wave propagation over flood plains. *14th Cong. Int. Ass. hydraul. Res. 5,* Paper 316, 217-219.

Shaw, T. L., Huntington, S. W. 1973. The use of off-stream storage to reduce flood peaks in rivers. *23rd Int. navigatn. Congress (PIANC), Ottawa.*

Huntington, S. W., Shaw, T. L. 1973. Application of mathematical models to flow in small streams. *15th Cong. Int. Ass. hydraul, Res., 5* Paper, 4,7, 149-152.

Effect of urban development on catchment hydrology
Dr T. L. Shaw (Dr G. R. Thorpe (1969-73),
H. S. Wheater, R. Marshall, N. P. Markovits) 1969-76

Extensive field monitoring and laboratory catchment studies provide data for computer simulation of the processes of rainfall and run-off for assessment of the various effects of urbanisation on catchment hydrology, especially flooding. The mathematical streamflow synthesis is a finite difference solution of the equations of flow within a matrix of cuboid elements into which a catchment is divided.

Shaw, T. L. 1973. Requirements of drainage design for large urban areas; Thorpe, G. R. Forecasting run-off from storms moving over partially urbanised catchments. *CIRIA Colloquium on Rainfall, Runoff and Surface Water Drainage of Urban Catchments, Bristol.*

Shaw, T. L., Waller, R. S. 1973. The Gloucester joint surface water study. *J. Instn. municip. Engrs., 100,* 120-123.

Smith, A. R., Waller, R. S., Wheater, H. S., Shaw, T. L. 1973. A study of the unsaturated zone in the hydrological balance of a small catchment. *15th Cong. Int. Ass. hydraul. Res.*

Marshall, R., Shaw, T. L. 1973. An approach to the stochastic modelling of rainfall. *Am. geophysical Union, Fall Meeting.*

Shaw, T. L., Marshall, R., Wheater, H. S., Markovits, N. P. 1974. Field monitoring equipment for hydrological studies in a small catchment. *Int. Ass. scient. Hydrol. Symposia on Flash Floods, Paris.*

Department of Geography (BTLGG)

Estimation and prediction of rainfall from weather satellite data

Dr E. C. Barrett (R. Harris, R. Gurney) 1969-

Satellite photographic and improved data are being analysed for the indications their cloud contents give of precipitation distribution and intensities. Statistical techniques are being used to correlate satellite rainfall indications with rainfall received at the ground in the preparation of nomograms for operational use.

Barrett, E. C. 1970. The estimation of monthly rainfall from satellite data. *Monthly Weather Review, 98,* 198-205.

Barrett, E. C. 1973. Forecasting daily rainfall from satellite data. *Monthly Weather Review, 101,* 215-222.

Barrett, E. C. 1973. *Satellite data in rainfall and catchment research.* British Ass. adv. Sci., Section E, Canterbury, 16pp.

Barrett, E. C. 1974. Daily and monthly estimates of rainfall from satellite data. In: *Environmental remote sensing: applications and achievements* (Eds. E. C. Barrett, L. F. Curtis). Edward Arnold, London, 241-66.

Patterned ground in the environs of the Severn Estuary

Dr L. F. Curtis (Mrs K. M. Trudgill, A. Tilstone) 1970-73

The relationship between shrinkage patterns in peats and clays and soil moisture, fluctuations in groundwater levels and other variables has been investigated in the field and laboratory. Predictive models of shrinkage of the landscape with falling water tables are under study.

Limestone hydrology and water tracing

D. I. Smith (Dr T. C. Atkinson (1970-73), P. L. Smart (1972-75)) 1964-

The form and variations of flow associated with various limestone lithologies, including Carboniferous Limestone, Chalk and Jurassic Oolites, are studied by use of fluorescent dyes and particulate tracers.

Atkinson, T. C. 1971. The dangers of pollution of limestone aquifers with special reference to the Mendip Hills, Somerset. *Proc. Univ. Bristol spel. Soc. 12, 3,* 281-90.

Smith, D. I., Atkinson, T. C. 1973. Underground flow rates in cavernous and fissured limestones. *Proc. wat. res. Ass. Conf. Groundwater Pollution, 3,* 71-75.

Atkinson, T. C., Smith, D. I., Lavis, J. J., Whitaker, R. J. 1973. Experiments in tracing ground waters in limestone. *J. Hydrol., 19,* 1-27, 323-49.

Smith, D. I., Newson, M. D. 1974. The dynamics of mechanical erosion in limestone catchments on the Mendip Hills, Somerset. *Inst. Brit. Geogr. spec. Pubn., 6,* 155-167.

Smith, D. I., Drew, D. P. 1974. Eds. *Limestone and caves of the Mendip Hills,* 425pp. David and Charles.

Atkinson, T. C., Drew, D. P. 1974. *Underground drainage of limestone catchments in the Mendip Hills. Inst. Brit. Geogr. spec. Publ., 6,* 87-106.

Atkinson, T. C., Smith, D. I. (In press). Rapid groundwater flow in fissures in the Chalk – an example from south Hampshire. *Quart. J. engng. Geol.*

Smith, D. I., Atkinson, T. C., Drew, D. P. (In press). Hydrology of limestone terrains. In: *The sciences of speleology,* David and Charles.

Smith, D. I., Atkinson, T. C. 1975. (In press). Process, landforms and climate in limestone regions. In: *Geomorphology and Climate* (Ed. E. Derbyshire). Wiley.

Suspended sediments in tributary rivers of the Severn Estuary
D. E. Smith (Mrs Rosemary Brookes) 1971-74

Suspended sediment concentrations of the rivers Severn, Wye, Usk and Bristol Avon were measured and combined with flow duration data to obtain an estimate of the quantity of suspended sediment discharged into the Severn Estuary. Detailed studies were also undertaken on 10 sub-catchments within the Bristol Avon watershed. Solutes were also measured.

Hydrological processes affecting downstream changes in water quality
Dr M. G. Anderson, D. I. Smith (T. Burt) 1974-

Analysis of the downstream changes in discharge water chemistry and particulate load in a 36 km² basin in North Somerset.

Deep quarrying and groundwater movement in the Carboniferous Limestone of the Bristol region
D. I. Smith with Dr B. P. J. Williams, Department of Geology (D. C. Holmes) 1973-

The expansion of the limestone quarrying industry has resulted in an increase in the depth of working so that soon workings will become sub-water table. The hydrological characteristics of the limestones will be assessed to predict the magnitude of the pumping required and the effects on the water resources of the area.

Atkinson, T. C., Bradshaw, R., Smith, D. I. 1973. *Quarrying in Somerset, Supplement No. 1.* 58pp, Somerset Cty Council.

Department of Geology (BTLGL)

Groundwater studies of East Midland aquifers
Dr B. P. J. Williams 1968-

Groundwater resources within the Lincolnshire Limestone aquifer are being evaluated and relationships between the sedimentological and hydrological properties of the Bunter Sandstone aquifer are being investigated.

Williams, B. P. J., Downing, R. A., Lovelock, P. E. R. 1972. Aquifer properties of the Bunter Sandstone, Nottinghamshire, England. *Proc. 24th Int. geol. Cong. Montreal, Section 11 – Hydrogeology*, 7 pp.

Hydrogeology of Jurassic limestones of the Cotswolds between Bath and Tetbury
Dr B. P. J. Williams (J. Bromley) 1972-75

Groundwater conditions including flow regimes, aquifer properties and aquiclude leakage are evaluated through detailed analyses of pumping tests. The relationships between groundwater chemistry and lithofacies variation is being investigated.

CAMBRIDGE

Department of Geography (CAMGG)

River discharge characteristics and associated basin forms
Dr R. J. Chorley (M. G. Anderson) 1970-74

Relationships between basin form and river discharge were investigated by time series models fitted to discharge data from twenty basins in Britain and a model capable of reproducing the three-dimensional form of drainage basins is proposed. These models demonstrate the dominant control that basin form exerts upon the Markov component in the streamflow series. The 'at-a-station' analysis is complemented by downstream analysis of streamflow in the Snowy Mountains of Australia.

Anderson, M. G. 1973. Measure of three-dimensional drainage basin form. *Wat. resour. Res. Wash.*, *9*, 378-383.

Theoretical models of river channel pattern
Dr R. J. Chorley (Dr R. I. Ferguson) 1968-72

A conceptual ratio between stream power and bank resistance is used to discriminate among the continuum of sinuous channels. Possible forms of differential equations relating sinuosity and curvature to bank erosion were shown to be equivalent to an inertial-type dependence on angular deviation from the downslope direction, subject to a curvature limit possibly related to cross-sectional form. A computer model for meander paths was developed and applied to some British rivers.

Ferguson, R. I. 1973. Sinuosity of supraglacial streams. *Bull. geol. Soc. Am.*, *84*, 251-256.

Ferguson, R. I. 1973. Regular meander path models. *Wat. resour. Res. Wash.*, *9*, 1079-1086.

Ferguson, R. I. 1973. Channel pattern and sediment type. *Area*, *5*, 38-41.

Law and order in stream networks
Dr R. J. Chorley (R. S. Jarvis) 1970-74

A computer-based analysis of the topology and geometry of stream channel networks based on the semi-automatic generation of data by means of digitizer.

Jarvis, R. S. 1972. New measure of the topologic structure of dendritic drainage networks. *Wat. resour. Res. Wash., 8,* 1265-1271.

Factors influencing the hydraulic geometry of river channels
Dr R. J. Chorley (Dr R. D. Hey) 1965-72

A conceptual model was established to identify the nature of the interaction between the geometry of the system and the hydraulic and surficial variables. The dependent variables included wetted perimeter, hydraulic radius, maximum depth, channel slope, velocity and sinuosity. The independent variables (discharge, sediment discharge, type of bed and bank material) were identified after obtaining expressions for the six governing equations which control these channel adjustments. These equations involve continuity, flow resistance, bed load, bank competence (two banks) and sinuosity. The model was checked using data collected from 76 sites on the rivers Wye, Severn and Tweed.

Initial development of stream channel networks, with special reference to the role of soil piping
Dr R. J. Chorley (J. A. A. Jones) 1967-73

The factors controlling the formation and distribution of soil piping are investigated to establish how common the process may be and how the structure of the pipe network might be transformed into a stream channel network. The structure of percoline networks is also examined. Piping is found to be widespread in the British upland areas and both pipe and percoline networks tend to exhibit features comparable with channels but on a different scale.

Jones, J. A. A. 1971. Soil piping and stream channel initiation. *Wat. resour. Res. Wash., 7,* 602-610.

Hydraulic geometry of channels of variable roughness
Dr R. J. Chorley (K. S. Richards) 1970-74

A hydraulic geometry model is investigated at several scales to generalise the model and improve its descriptive and explanatory power. At-a-station hydraulic geometry was studied using a power function model, and applied to English river data. The implications for regime and minimum variance theories were considered. Field data for the River Fowey demonstrated the effect of the bed profile variations on the hydraulic geometry of pool and riffle sections. Downstream hydraulic geometry relationships were investigated using measured bankfull cross-sections and samples of bed and bank material.

Richards, K. S. 1972. Meanders and valley slope. *Area, 4,* 288-290.

Richards, K. S. 1973. Hydraulic geometry and channel roughness – A non-linear system. *Am. J. Sci., 173,* 877-896.

River networks in South-West England
Dr R. J. Chorley (A. Werritty) 1969-74

The topological and geometrical characteristics of stream networks in Devon were analysed with particular reference to structural control. The work is based on field, air photo and map sources, using digitizer and computer.

Werritty, A. 1972. Accuracy of stream link lengths derived from maps. *Wat. resour. Res. Wash., 8,* 1255-1264.

Werritty, A. 1972. The topology of stream networks. In: *Spatial Analysis in Geomorphology* (Ed: R. J. Chorley), Methuen, 167-196.

Department of Physics (CAMPH)

Friction, deformation and recrystallisation of single crystals of ice IH under stress
Prof. D. Tabor (Prof. E. L. Offenbacher,
Mr I. C. Roselman) 1971-

Recrystallisation during sliding is examined by sliding a shallow cone of a single crystal of ice, supported on a glass substrate, over the surface of another glass plate. The interface shows that appreciable recrystallation does occur but that this takes place under normal loading and is not increased by the sliding process. The orientation is not with the basal plane strictly parallel to the direction of sliding but at an average angle of 35°. This may still be sufficient to explain the frictional behaviour in terms of an easy creep orientation.

Offenbacher, E. L., Roselman, I. C. 1971. Hardness anistropy of single crystals of ice IH. *Nature, phys. Sci. Lond., 234, 49,* 112-113.

Offenbacher, E. L., Roselman, I. C., Tabor, D. 1972. Friction, deformation and recrystallisation of single crystals of ice IH under stress. *Proc. Ottawa Symposium on the Physics and Chemistry of Ice, 80.*

Influence of limited solubility on the electrical and mechanical properties of ice
Dr J. C. F. Walker with Dr J. G. Paren
(Scott Polar Research Institute) 1971-

The influence of impurities in natural ice when the concentration exceeds the small but finite solubility limit is examined. At low temperatures the hydrated salts have negligible dielectric loss. Above –10° temperature dependent changes in permittivity can be identified with the equilibrium component of liquid brine. This second phase has analogous effects on the creep properties of polycrystalline ice.

Paren, J. G., Walker, J. C. F. 1971. Influence of limited solubility on the electrical and mechanical properties of ice. *Nature, phys. Sci. Lond., 230,* 91-92.

CITY

Department of Civil Engineering (CTYCE)

Computer simulation of water resource systems
Dr W. J. Morris (with Dr N. W. Morris)
(M. McGovern, Dr F. Saba) 1966-

Computer simulation of complete catchments with the aid of analog, digital and hybrid computers for more efficient design, operation and management of water resource systems (both above and below ground).

Morris, W. J. 1970. The synthesis of hydrologic systems on the analog and hybrid computers. *Proc. 2nd Int. Sem. for hydrol. Prof., Nat. sci. Found. Adv. Sci.* Utah State University, U.S.A. 275-301.

Morris, W. J., Morgan, N. W. 1971. Hybrid computer analysis of a combined surface water/groundwater system. *Proc. int. ass. scient. Hydrol. Int. Symp. on Mathematical Models in Hydrology, Warsaw, Poland.*

Morris, W. J. 1972. Optimization of water resource systems. *Proc. hybrid optimlzation Symp. 1972.* 1-12.

Morris, W. J. 1972. Parameter identification of water resource systems. *Trans. Computers in Education Div. of Am. Soc. for Engng. Educat., 5, 6,* 85-90.

Morris, W. J., McGovern, M. C. 1973. Hybrid computer optimisation of a catchment simulation model. *Proc. int. Symp. on Optimisation in civ. Engng., Univ. Liverpool,* 1973.

Thermodynamics of soil moisture and minerals
J. D. Coleman (R. Farrell, C. Begg) 1945-

Free energy and entropy studies are made of soil-moisture distribution and mineral stability including dehydration, rehydration and dehydroxylation reactions studies in Stanton TR1 thermobalance. Phases produced at various temperatures are identified by X-ray powder camera. The aim is to predict reaction temperatures using thermodynamic tables for minerals, and especially to find how soils and minerals differ from pure laboratory preparations, and from one to another, in water loss and gain.

Coleman, J. D. 1974. Total heat, free energy and entropy in classification of natural moisture and moisture processes. *Ministry of Agriculture, Fisheries and Food, Tech. Bull., 29,* 208-210, H.M.S.O., London.

Coleman, J. D. 1970. Distribution of water and moisture in soil media. *Civ. Engng. Publ. Works Rev. Lond., 65,* 771, 1169-1170.

Coleman, J. D. 1972. Evaporation pan in Central London. *Weather, Lond., 27,* 2, 93-94.

DUNDEE

Department of Civil Engineering (DUNCE)

Water quality on peat catchments
W. McNicoll with Dr H. A. P. Ingram. Dept. of Biological Sciences (P. J. Mott, D. J. Tervet) 1970-73

An experimental catchment includes a mire of the raised bog type. Elements of a water balance and mineral nutrient budget of this mire were derived from general and local measurements involving a stream-gauging network and automatic water sampling. The role of peat and mire vegetation in the budget is being assessed in relation to plant productivity and peat accretion.

Department of Geology (DUNGL)

Bed load and suspended sediment budgets in the Tay Basin, Scotland
Dr J. McManus (N. A. A. Ansari) 1973-81

Bed and suspended sediment loads are being measured for a range of flow conditions at established gauging stations throughout the Tay drainage system in catchments of a variety of geological backgrounds. The results predicted by a range of equations are compared with measured loads. The importance of entrapment in locks will be determined by bathymetry after a 70-year interval.

Mechanical Engineering (DUNME)

Hydraulic conductivity of humified peats
Dr J. M. B. Brown with Dr H. A. P. Ingram, Dept. of Biological Science (J. Waine) 1971-74

The physics of water transmission through humified peats are reappraised to establish the nature of the non-linear relationship between flux and potential gradient. Laboratory studies employing themostatted and chemically inert permeameters are combined with variable head seepage tube tests in the field in which the initial level displacements are established initially instantaneously.

EAST ANGLIA

School of Environmental Sciences (Climatic Research Unit) (EANCR)

Modelling rainfall in South East England
Mrs B. M. Gray 1972-75

Between 5 and 18 stations were used to set up areal monthly average rainfall. The best fit distribution was an incomplete gamma distribution with $r = 3{\cdot}41$, $\lambda = 1.50$.

School of Environmental Sciences (EANEV)

A deterministic, spatially distributed model of catchment hydrology
Prof. K. M. Clayton (K. J. Beven) 1971-74

A physically-based model of catchment hydrology has been developed using finite element and finite difference methods to solve the partial differential equations of transient water flow in soils and open channels. The model has been applied to real drainage basins using only data that can be measured or estimated directly and is suitable for use in the short term forecasting problem in ungauged basins.

Variations in the major dissolved constituents of certain East Anglian rivers
Dr P. S. Liss, Prof. K. M. Clayton (Dr A. M. C. Edwards) 1969-71

The origins and variations in the major cation and anion concentrations were investigated by weekly sampling at discharge gauging stations and by analysis of precipitation.

Edwards, A. M. C. 1973. Dissolved load and tentative solute budgets of some Norfolk catchments. *J. Hydrol.*, *18*, 201-217.

Edwards, A. M. C. 1973. The variations of dissolved constituents with discharge in some Norfolk rivers. *J. Hydrol.*, *18*, 219-242.

Edwards, A. M. C., Liss, P. S. 1973. Evidence for buffering of dissolved silicon in fresh waters. *Nature, Lond.*, *243*, 341-342.

Some effects of water on slope recession
Prof. A. Young (M. J. Day) 1971-74

Precipitation, run-off and through-flow are metered and analysed for dissolved solids (Ca, K, Mg, Na, Fe, Mn, Zn, XI, Si, Al, SO_4) and suspended sediment at three points on a convex, 20-30° slope near Aberystwyth.

Use of fluorescent dyes for tracing movement of soil moisture

Dr T. C. Atkinson 1971-

Pilot experiments were carried out in order to test the use of fluorescent dyes in tracing soil moisture flows before application in tracing pipeflow and pipe networks in steep catchments; downslope throughflow on steep slopes, and unsaturated infiltration in forest floors. Detailed studies of infiltration at Thetford, Norfolk, especially of small scale variability of flow through nearly homogenous sand are in co-operation with the Institute of Hydrology.

Atkinson, T. C. (In press). Measurement of subsurface flow on hillslopes. Chap. 3 in *'Hillslope Hydrology'*, (Ed. M. J. Kirkby). J. Wiley.

Tracer studies in the Norfolk Broads

Dr T. C. Atkinson 1974-

Fluorescent dyes are being used as tracers to determine the mixing and dispersion of inlets of pumped water from land drainage schemes in the Norfolk Broads. The information obtained provides a physical background to the aquatic ecology and may prove useful in dealing with future outbreaks of the toxic algae, *Prymnesium*.

Effects of drainage network geometry on stream hydrograph form

Dr A. Calver 1972-

Hillslope water contributions to the stream channel are routed through the drainage network to determine the stream outflow hydrograph for a range of network geometries.

Calver, A., Kirkby, M. J., Weyman, D. R. 1972. Modelling hillslope and channel flows. In: *Spatial Analysis in Geomorphology*, (Ed. R. J. Chorley). Methuen, 393p.

Relationships between soil moisture retention characteristics and irrigation requirements

D. L. Dent, Dr R. D. Hey, with W. Corbett
(Soil Survey of England and Wales) 1974-77

Soil moisture retention profiles will be evaluated as guidelines for irrigation scheduling by determination of soil moisture retention profiles for major soil series in East Anglia and by experimental work on relationships between texture, structure, management and moisture retention characteristics.

Water and suspended sediment movement in Breydon Water, East Anglia

Dr T. G. Harvey (P. L. Firstbrook) 1972-75

The movement of water and suspended sediment is being examined in the estuary during complete tidal cycles. In addition, water movement across the large expanses of mud-flats, and within the channels which drain the flats, is under examination.

Alluvial channels: processes and form

Dr R. D. Hey 1972-

Continuation of research on the simulation of erosional and depositional activity in rivers. (see CAMGG).

Hey, R. D. (In press). Prediction and effects of flooding in alluvial systems. *Quart. J. geol. Soc. London.*

Hey, R. D., Thorne, C. R. (In press). Secondary flows in river channels. *Nature, Lond.*

Flood alleviation feasibility studies
Dr R. D. Hey, Dr B. G. Clarke 1973-74

Application of modified cost-benefit analysis to establish optimum strategy for a flood alleviation scheme at Hereford, River Wye.

Factors controlling the concentration of dissolved phosphorus in natural waters
Dr P. S. Liss (T. P. Coles)

In order to ascertain whether mineral-water reactions are important in controlling dissolved phosphorus levels in natural waters, both river and estuarine waters are being studied and, where possible, field results are to be investigated through laboratory simulation experiments.

EDINBURGH

Department of Forestry and Natural Resources (EDGFN)

Response of trees to windspeed and humidity
Dr J. Grace (J. Wilson) 1973-76

A controlled environment wind tunnel is being used to study the influence of windspeed and humidity on diffusive resistance and water relations in a range of tree species. Emphasis will be on the importance of wind in controlling the growth and ecological characteristics of trees.

Studies of land use adjustment to flood risk in Southern Scotland
Dr D. C. Ledger (A. T. McDonald) 1968-74

Models based on hydrological and topographical data to predict flood-prone areas have been used with regression models to assess physical damage and associated monetary losses caused by flooding of agricultural land in the Nith Valley. An evaluation has been made of flood risk on agricultural practices in areas both with and without levee protection.

Erosion and sediment-yield studies in southern Scotland
Dr D. C. Ledger, Dr J. P. B. Lovell 1970-

Periodic surveys of gullies, stream channels and reservoirs have been made to determine rates of erosion and sedimentation in upland areas in southern Scotland. More detailed studies of erosion processes and their relation to land use are planned.

Lovell, J. P. B., Ledger, D. C., Davies, I. M., Tipper, J. C. 1973. Rate of sedimentation in the North Esk Reservoir, Midlothian. *Scot. J. Geol., 9,* (1), 57-61.

Ledger, D. C., Lovell, J. P. B., McDonald, A. T. 1974. Sediment yield studies in upland catchments areas in South-East Scotland. *J. appl. Ecol., 11.* (1), 201-206.

Influences on the growth of Sitka spruce
Dr D. C. Malcolm (Dr I. K. Bradbury, B. C. Y. Freezaillah) 1972-76

Tree seedlings (Sitka spruce) of various provenances are being grown in soil media of known fertility at 5 stations on a hill slope from 220m to 475m. Climatic factors are recorded at each station and used to study tree growth responses as well as record variation with elevation. Nutrient loss in leachates is monitored and the effect of differential nutrition on water use is being investigated.

Malcolm, D. C., Freezaillah, B. C. Y. 1973. The interactions of climate and phosphorus fertilisation on the growth of Sitka spruce (*Picea sitchensis* Bong., Carr.) seedlings. *FAO/IUFRO int. Symp. forest Fertilization, Paris, FAO/UFRO(F) 73/32.*

Hydrological characteristics of fish spawning streams
Dr D. H. Mills (Miss B. D. Smith) 1968-75

Included in an ecological study of a number of fish species in the Tweed river system is an assessment of physical and hydrological characteristics of certain spawning streams and reasons for the preference of certain streams by salmon, sea trout and brown trout.

Mills, D. H. 1973. Preliminary assessment of the characteristics of spawning tributaries of the River Tweed with a view to management. *International Atlantic Salmon Symposium*, International Atlantic Salmon Foundation.

Department of Meteorology (EDGMT)

Physiological and synoptic-scale influences in diurnal humidity variations over vegetation
Dr A. S. Thom (J. C. Curran) 1972-

The research forms part of the Institute of Hydrology's study of evaporation from forest, based on field observations near Thetford, Norfolk.

Stewart, J. B., Thom, A. S. 1973. Energy budgets in pine forest. *Quart. J. roy. met. Soc., 99,* 154-170.

EXETER

Department of Geography (EXEGG)

Drainage basin changes and their hydrological significance
Dr K. J. Gregory 1970-

Changes of drainage networks are studied in relation to world values of drainage density to provide an improved understanding of the significance of the drainage network for water and sediment production. The magnitude of channel change is analysed in relation to the drainage net, indicated by measurements of sediment movement, by monumented cross sections, and by reference to situations in basins subjected to substantial hydrologic change including reservoir construction.

Gregory, K. J. 1971. Drainage density changes in South West England. In: *Exeter Essays in Geography* (Eds. K. J. Gregory, W. L. D. Ravenhill), 33-53.

Gregory, K. J. 1971. River networks in Devon. *Proc. Torquay nat. Hist. Soc., 16* (*1*), 4-11.

Gregory, K. J., Walling, D. E. 1971. Field measurements in the drainage basin. *Geography, 56,* 227-92.

Gregory, K. J., Walling, D. E. 1973. *Drainage basin form and process: a geomorphological approach* Arnold. 456p.

Gregory, K. J., Walling, D. E. (Eds.). 1974. *Fluvial processes in instrumented watersheds. Inst. Brit. Geogr. spec. Pubn., 6, 196p.*

Gregory, K. J. (In press). Drainage networks and climate. In: *Climate and Landforms*, J. Wiley.

Gregory, K. J., Park, C. C. (In press). Adjustment of river channel capacity downstream from a reservoir. *Wat. resour. Res. Wash.*

Flood plain and valley floor development in the Upper Exe basin
Dr K. J. Gregory (B. Cant) 1971-74

The later stages of valley development in the Upper Exe and Barle valleys are being investigated using detailed mapping of present flood plain morphology, augering of flood plain deposits, and palaeohydrologic analysis of buried channels by comparison with contemporary channels and with results from other humid and periglacial areas.

Cant, B. (In press). Aspects of valley floor development in the Upper Exe Basin. *Trans. Devon. Ass.*

Physical land classification in South West England
Dr K. J. Gregory (Dr V. Gardiner) 1969-72

Mapping of drainage density largely based upon 1:25,000 maps was used as a basis for the production of maps of basin character. A technique of mapping estimated drainage density based upon grid squares has been developed, applied to South West England and parts of Uganda, and analysed in relation to other basin characteristics.

Gardiner, V. 1971. A drainage density map of Dartmoor. *Trans. Devon. Assoc., 103,* 167-180.

Gardiner, V. 1974. Univariate distributional characteristics of some morphometric variables. *Geograf. Annal., Ser. A, 54A,* 147-153.

Gardiner, V. (In press). Land form and land classification in Northwest Devon. *Trans. Devon. Assoc.,* 106.

Gardiner, V. (In press). A photo-mechanical technique for the production of drainage density maps. *Cartog. J., 11.*

Gardiner, V., Rhind, D. 1974. The creation of slope maps by a photo-mechanical technique. *Area, 6.,* 14-21.

Gardiner, V. (In press) Drainage basin morphometry. *Brit. geomorphol. Res. Grp. tech. Bull., 14.*

Gregory, K. J., Gardiner, V. (In press). Drainage density and climate. *Z. geomorphol.*

Temporal changes of channel pattern
Dr K. J. Gregory (Miss J. C. Hooke) 1973-

The distribution and magnitude of changes of channel pattern in Devon since the early nineteenth century is studied by reference to surveys, maps and aerial survey of different dates. The mechanism of short-term change is being investigated along 8 instrumented and monumented channel reaches. Analysis is directed towards the controls of changes of channel pattern at both scales.

Controls of stream channel geometry
Dr K. J. Gregory (C. C. Park) 1972-

The size, form and spatial variations of stream channels in Devon is being investigated to provide information upon the controlling factors and upon the relationships of channel character to stream discharge. Measurements are made along major rivers, the pattern of downstream variation is characterised and is related to hydrological parameters.

Spatial variations in rainfall input to a small instrumented catchment
Dr K. J. Gregory (Mrs E. J. Richardson) 1969-72

A network of 45 totaliser gauges and two autographic gauges was used to provide information on the variation of precipitation over a 3.9km^2 instrumented catchment. Patterns of variation were classified according to storm characteristics and the magnitude of variation analysed in relation to a dense network including ground level gauges.

Effects of building activity and urbanisation upon streamflow, sediment and solute yield
Dr K. J. Gregory, Dr D. E. Walling 1968-

Detailed changes in a small (0.25km^2) catchment are being monitored during the progress of building activity and the project, which started when the catchment was under agricultural use, is designed to continue until urbanisation is complete. Continuous records of precipitation and streamflow are maintained, other climatic parameters are monitored locally, and water samples obtained automatically for individual storm hydrographs are analysed for suspended sediment and solute content.

Walling, D. E., Gregory, K. J. 1970. The measurement of the effects of building construction on drainage basin dynamics. *J. Hydrol.*, *11*, 129-144.

Walling, D. E. 1970. Sediment production in a small catchment before urbanisation. *Area*, *3*, 64-5.

Gregory, K. J. 1970. Rainfall-streamflow relationships for an instrumented catchment. *Area*, *3*, 65-6.

Gregory, K. J. 1974. Streamflow and building activity. *Inst. Brit. Geogr. spec. Publ.*, *6*, 107-122.

Walling, D. E. (In press). Suspended sediment production and building activity in a small British catchment. *Int. Ass. scient. Hydrol.* Paris Symp.

Runoff production and the influence of catchment characteristics
Dr D. E. Walling 1967-71

The processes of runoff production were examined in five small instrumented catchments in south-east Devon. The variable source area and partial-area concepts of storm runoff production were found to be fundamental in understanding catchment response. Contrasts in streamflow response between the five catchments were evaluated and related to catchment characteristics, particularly geology and land use.

Walling, D. E. 1971. Streamflow from instrumented catchments in south-east Devon. In: *Exeter Essays in Geography* (Eds: K. J. Gregory, W. C. D. Ravenhill), 55-81.

Gregory, K. J., Walling, D. E. 1971. Field measurements in the drainage basin. *Geography, 56,* 277-92.

Sediment and solute dynamics of Devon catchments
Dr D. E. Walling 1968-

The processes of sediment and solute production in Devon catchments are investigated by detailed study of seven small instrumented catchments on contrasting terrain types and by a large scale investigation on the River Exe. Mathematical models of sediment and solute production are being developed, and yields and denudation rates are being evaluated.

Walling, D. E. 1971. Sediment dynamics of small instrumented catchments in south-east Devon. *Trans. Devon. Ass., 103,* 147-65.

Walling, D. E., Teed, A. 1971. A simple pumping sampler for research into suspended sediment transport from small catchments. *J. Hydrol., 13,* 325-37.

Gregory, K. J., Walling, D. E. 1973. *Drainage basin form and process: a geomorphological approach.* Arnold. 456p.

Gregory, K. J., Walling, D. E. 1973. Fluvial processes in small instrumented watersheds in the British Isles. *Area, 5.* (*4*), 297-302.

Walling, D. E. 1974. Suspended sediment and solute yields from a small catchment prior to urbanisation. *Inst. Brit. Geogr. spec. Pub., 6,* 167-92.

Walling, D. E. (In press). Solute variations in small catchment streams: some comments. *Trans. Inst. Brit. Geogr.*

Hydrology of the catchment area of Slapton Ley
Dr D. E. Walling with R. P. Troake
(Slapton Ley Field Centre) 1970-

As part of a wider study of the freshwater biology of Slapton Ley, work is in progress to document the runoff and sediment and solute dynamics of the streams draining into the lake. Particular attention is being given to the replacement time of the lake and the input of nutrients and sediment.

Troake, R. P., Walling, D. E. 1973. The hydrology of the Slapton Wood stream. *Field Studies, 3.* (5), 719-740.

Influence of catchment characteristics on low flows in Southwest England
Dr D. E. Walling (T. J. Browne) 1972-

Characteristics of low flows in Devon streams, including frequency plots and recession curves are being analysed by relating spatial contrasts in low flow response to catchment characteristics, particularly geology. The investigation involves both existing hydrometric data and a programme of spot gauging. Maps of various minimum flow indices are being constructed.

Evaluation and mapping of the water resource characteristics of Dartmoor
Dr D. E. Walling (J. Buckett) 1971-74

Spatial and temporal extrapolation of the limited available hydrometric data is being undertaken to provide a regional appraisal of the hydrological characteristics of Dartmoor and the surrounding areas. Methods for data infilling and for extension of runoff and rainfall records using stochastic generation techniques have been developed. Hydrological maps of the region have been produced.

Interrelationship of streamflow and water quality in a small catchment
Dr D. E. Walling (I. D. L. Foster) 1973-

An intensive scheme of instrumentation has been installed in a small (1.4km^2) catchment in order to evaluate the interrelationships of runoff and water quality dynamics. The concentrations of individual cations and anions are being monitored using automatic sampling apparatus. Mathematical models of runoff and solute transport are being developed.

Application of the grid square technique to hydrological mapping
Dr D. E. Walling (Miss A. M. Foyster) 1970-73

The grid square technique has been applied to the evaluation and mapping of water balance components within an area of Southwest England. Computer routines for storage, retrieval, processing and mapping of physiographic and hydrological data on a grid square basis have been developed.

Foyster, A. M. 1973. Application of the grid square technique to mapping of evapotranspiration. *J. Hydrol., 19,* 205-26.

Spatial and temporal variations in solute transport within the Exe catchment
Dr D. E. Walling (B. W. Webb) 1971-

Intensive spatial sampling (> 500 sites) and frequent sampling at major gauging stations are being employed to document spatial and temporal variation in the concentrations of selected cations and anions in the streams of the Exe basin (1462km^2) and particularly the Upper Exe catchment (613km^2). The results are being interpreted in terms of control by catchment characteristics and streamflow dynamics and are being used to develop a mathematical model of solute transport in the River Exe.

Walling, D. E., Webb, B. W. (In press). Spatial variation of river water quality. *Trans. Inst. Brit. Geogr.*

GLASGOW

Department of Civil Engineering (GLACE)

Rainfall and run-off correlations for small catchments
Dr J. G. Herbertson, with Dundee College of Technology 1970-

Two catchments, one highland and one semi-coastal, are being instrumented to provide continuous rainfall and streamflow records. The object is to obtain a detailed picture of the run-off from individual sections of the catchments and to use the data to calibrate run-off models.

HULL

Department of Plant Biology (HULPB)

Water level fluctuation in peat in relation to rainfall
Dr D. J. Boatman (P. Hulme) 1970-73

The aim was to explain the origin and development of deep pools which occur on all bogs of the series in the

Silver Flowe nature reserve. Water level was recorded periodically in tube wells and Munro recorders have been used to obtain continuous records of the movement of water level in pools and peat. Rainfall has also been measured.

Boatman, D. J., Tomlinson, R. 1973. The Silver Flowe I. Some structural and hydrological features of Brishie bog and their bearing on pool formation. *J. Ecol., 61*, 653-666.

Department of Geography (HULGG)

Throughflow on hillslopes in the North Yorkshire moors
Dr R. R. Arnett 1969-72

Analysis of weekly throughflow samples from 15 sites, together with other climatological, pedological and fluvial information.

Arnett, R. R. 1972. The field measurement of lateral soil water movement. *Rev. Geomorphol. dynamique, 11*.

Arnett, R. R. 1974. Environmental factors affecting the speed and volume of topsoil interflow. *Inst. Brit. Geogr. spec. Publ., 6*, 7-22.

Factors controlling sediment production in the North Yorkshire Moors
Dr R. R. Arnett 1974-76

Establishment of sediment/discharge rating curves for 20 catchments, relating sediment yields to lithology, soils and land-use variables.

Karst water studies in the Central Pennines
Dr A. F. Pitty (Dr J. L. Ternan) 1967-74

Analyses have been made of water samples collected at fortnightly intervals from a large number of points around Malham Moor. Qualitative analysis appears to confirm a positive correlation with antecedent temperature for several springs.

Catchwater catchment experiment
Dr R. C. Ward (Dr M. Bonell, Dr A. C. Imeson, R. K. Pegg, Dr B. L. Oyebande, I. Roper, J. D. Tandy, Miss E. Wells, Dr M. Visvalingam) 1965-85

To study its hydrology, a small boulder-clay catchment has been comprehensively instrumented to measure a wide range of variables.

Ward, R. C. 1971. Small watershed experiments: and appraisal of concepts and research developments. *Univ. Hull occl. paps. Geogr., 18*, 254 pp.

Ward, R. C. 1972. Checks on the water balance of a small catchment. *Nordic Hydrol., 3*, 44-63.

Ward, R. C. 1972. Estimating streamflow using Thornthwaite's climatic water balance. *Weather*, 73-84.

Ward, R. C., Imeson, A. C. 1972. The output of a lowland catchment. *J. Hydrol., 17*, 145-159.

Groundwater hydrology of a boulder-clay catchment
Dr R. C. Ward (Dr M. Bonell) 1967-71

A detailed network of more than 40 observation wells formed the basis of a study of the disposition and movement of groundwater at shallow depths within the boulder-clay.

Bonell, M. 1972. The application of the auger hole method in Holderness glacial drift. *J. Hydrol, 16,* 125-146.

Bonell, M. 1972. An assessment of possible factors contributing to well level fluctuations in Holderness boulder clay, East Yorkshire. *J. Hydrol., 16,* 361-368.

Infiltration and soil moisture movement in a boulder-clay plot
Dr R. C. Ward (Dr M. Visvalingam, J. D. Tandy) 1969-73

Soil moisture storage, movement and infiltration relationships were investigated using a neutron scatter probe, well-point techniques and ring infiltrometers.

Visvalingam, M., Tandy, J. D. 1972. The neutron method for measuring soil moisture content – a review. *J. soil Sci., 23,* 499-511.

Climate and the water balance in a clay catchment
Dr R. C. Ward (R. K. Pegg) 1966-74

Water balance variables measured in the Catchwater catchment have been statistically analysed to determine the magnitude of actual evapotranspiration losses.

Pegg, R. K., Ward, R. C. 1971. What happens to the rain? *Weather,* 88-97.

Pegg, R. K., Ward R. C. 1972. Evapotranspiration from a small clay catchment. *J. Hydrol., 15,* 149-165.

LANCASTER

Department of Environmental Sciences (LANEV)

Non-linear conceptual catchment model for isolated storm events
Prof. T. O'Donnell, with A. N. Mandeville, Institute of Hydrology 1971-74

A simple 4-parameter non-linear conceptual catchment model was developed and an effective optimisation fitting technique devised for modelling isolated storm events. When fitted, the model may be used for the design of flood events given design rainfalls and measures of initial catchment wetness. The project formed part of the Institute of Hydrology's Flood Studies work.

O'Donnell, T., Mandeville, A. N. 1975. *Floods Studies Report, 1,* Chap. 7, Natural Environment Research Council, London.

Flood design by unit hydrograph technique
Prof. T. O'Donnell, with M. J. Lowing, Institute of Hydrology 1971-74

An extensive study of storm events of UK Catchments and of procedures involved in unit hydrograph methodology has led to a flood-design method applicable to ungauged as well as gauged catchments in the UK. The project formed part of the Institute of Hydrology's Flood Studies work.

O'Donnell, T., Lowing, M. J. 1975. *Flood Studies Report, 1,* Chap. 6, Natural Environment Research Council, London.

Introduction of stochastic elements into deterministic catchment modelling
Prof. T. O'Donnell (D. Machado-Olive) 1973-76

In conventional optimisation procedures for fitting deterministic conceptual models to catchment records, each objective function evaluation is made with a trial set of parameter values held fixed for the whole of that function evaluation. The effect on fitting of allowing the parameter values to fluctuate stochastically *during* a function evaluation is to be explored.

Optimal design of water resources systems
Prof. T. O'Donnell, with J. C. Wilkinson,
Dept. of Operational Research
(D. K. Smith) 1973-75

The problems of optimising large simulation models of water supply systems are being studied to develop techniques for advising planners on the size of system that meets a vector of demands at minimal cost while satisfying reliability and other constraints.

Water chemistry and sediment in motorway drainage water
C. Patrick (D. P. H. Laxen) 1972-76

Drainage at outfalls of motorways drains is sampled to determine major and minor anions and cations, trace metals and some organic compounds. Sediment is analysed for size distribution, composition and adsorbed ions. Particular attention is paid to de-icing salt budget as a tracer. The aim is to establish any pollution hazard from road drainage.

Modelling of groundwater problems using digital computer methods
D. G. Read (L. J. Dyke) 1973-

Evaluation of various digital models of aquifer conditions.

Small-scale structure of rainstorms
D. G. Read, with Dr D. G. Felgate 1973
now Univ. Adelaide 1973-

Short-term raingauge records and the prevailing meteorological situation are analysed for a range of storm types. Correlation analysis of two-minute data from the Cardington rainfall experiment of the Meteorological Office is being examined initially, in an attempt to estimate the small-scale 'cell' sizes and their movement. The ultimate aim is to provide a quantitative (point/area) description of rainfall patterns observed on the ground connected with various meteorological conditions.

Felgate, D. G., Read, D. G. 1973. Statistical analysis of the small scale structure of storms observed by rain gauges. *Abstract, Symp. on Estimation of Storm Rainfall Patterns*, San Francisco, 1973. EOS, *Trans. Am. geophys. Union, 54, 11*, 1079.

LEEDS

Department of Geography (LDSGG)

Influence of soil and topography on contributing areas and stream hydrographs (BTLGG 1971-73)
Prof. M. J. Kirkby (Dr D. R. Weyman,
J. R. Wood) 1971-74

Field measurement and computer modelling have been carried out for three very small catchments (0.1km^2)

on Old Red Sandstone and Millstone Grit. Multiple stream gauging and surface piezometers were used to isolate sub-catchment hydrographs for comparison with observed areas of surface saturation. A model in which flows are routed through the soil following the topography in detail is being developed.

Actual evapotranspiration and soil moisture in the eastern Pennines
Dr J. C. Lockwood (K. Venkatasawmy) 1971-73

Soil moisture models were produced for two catchments and a lysimeter was used for estimates of actual evapo-transpiration.

Radiation budget of a small pasture catchment
Dr J. G. Lockwood (Miss L. R. Collier) 1972-75

Quantification of the radiation budget of a small pasture catchment near Harrogate, Yorkshire, to produce a physically-based mathematical model.

Collier, L. R., Lockwood, J. G. (In press). The estimation of solar radiation under cloudless skies with atmospheric dust. *Quart. J. roy. met. Soc.*

Topographic controls on the distribution of rainfall in the Central Pennines
Dr J. G. Lockwood (K. C. Goh) 1972-75

The effects of topography on annual, seasonal and individual storm rainfalls are studied.

Goh, K. C., Lockwood, J. G. (In press). An assessment of topographic controls on the distribution of rainfall in the Central Pennines. *Met. Mag. Lond.*

LIVERPOOL

Department of Geography (LIVGG)

Vegetation and hydrology of the Ainsdale sand dune nature reserve
R. K. Pegg, with J. F. Handley, Dept. of Botany (S. Edmondson) 1972-75

A drop in the dune watertable over the last few years is a cause of considerable concern. The possible reasons for this fall are being investigated using a partial water balance approach. A vegetational survey has been directed toward a consideration of the effect of water level changes on the associations in the reserve and their conservation.

LONDON

BIRKBECK COLLEGE

Department of Geography (BKLGG)

The importance of hydrometeorological and pedological factors in soil-water disposition
Dr I. Reid (supervised at Hull University by Dr A. F. Pitty 1968-71) 1969-72

Water-flux in the surface soil-layer, and hydrometeorological inputs and demands over a sampling year are studied under catchment conditions suffering strict control of lithological, pedological, slope-form and biotic variables. Seasonally-distinct soil-water disposition mechanisms were differentiated, while within-catchment soil-water response was shown to be functionally related to differences in evapotranspiration controlled by slope aspect.

Reid, I. 1973. The influence of slope orientation upon the soil moisture regime, and its hydrogeomorphological significance. *J. Hydrol.*, *19*, 309-321.

Reid, I. 1973. The influence of slope aspect on precipitation receipt. *Weather, 28,* 490-494.

Reid, I. 1973. The choice of assumed particle density values in soil tests. *Area, 5,* 10-12.

Changing surface-depression storage and infiltration opportunity in ploughlands
Dr I. Reid 1973-75

A micro-scale examination of soil-surface configuration change under the influence of weather and its pertinence to depression storage, infiltration, and surface-runoff in ploughlands.

Agricultural betterment resulting from land drainage and/or flood protection
Dr I. Reid (Miss S. H. Davies) 1974-

Detailed examination of soil-water disposition under tile-drainage and for flood protection with particular reference to clay soils. Besides the physical aspects of effective water evacuation, an examination will be made of the opportunity-costs of land-drainage improvement.

IMPERIAL COLLEGE

Department of Botany (ICLBT)

Modelling rainfall interception in forest canopies
Prof. A. J. Rutter, Prof. K. A. Kershaw (until 1970),
Dr A. J. Morton (Dr P. C. Robins) 1968-

A mathematical model has been derived combining environmental and stand parameters, from detailed observations in a stand of Corsican pine, for prediction in other stands of known characteristics. The model is being validated by comparison with published records of interception loss.

Rutter, A. J., Kershaw, K. A., Robins, P. C. and Morton, A. J. 1971. A predictive model of rainfall interception in forests. 1. Derivation of the model. *Agric. Meteorol., 9,* 367-384.

Rutter, A. J., Morton, A. J. and Robins, P. C. (In press). Generalisation of the model and comparison with observations in some coniferous and hardwood stands. *Agric. Meteorol.*

Robins, P. C. (In press). A method of measuring the aerodynamic resistance to the transport of water vapour from forest canopies. *J. appl. Ecol.*

Plant physiological influences on evaporation from forests
Prof. A. J. Rutter (Dr P. C. Robins,
Dr J. M. Roberts) 1968-74

In co-operation with the Institute of Hydrology, a quantitative assessment has been made of diurnal and seasonal changes in the stomatal resistance to transpiration of a Scots pine canopy, by use of a diffusion porometer, and this resistance has been related to irradiance leaf age and leaf water potential. The resistances to movement of water from soil, through the trees, to the foliage, and the effects of soil water potential and resistance in the plants on leaf potential and hence on transpiration have been assessed.

Department of Civil Engineering (ICLCE)

Time-variant linear catchment models
T. O'Donnell (now LANEV) 1969-72

The impulse response (I.U.H.) of time-variant versions of the linear reservoir and the linear channel were derived and used to study different paired combinations of the two basic time-variant components in series and in a cascade of time-variant linear reservoirs. The response of any time variant linear catchment model was investigated.

Mandeville, A. N., O'Donnell, T. 1973. Introduction of time variance to linear conceptual catchment models. *Wat. resour. Res. Wash., 9,* 2, 298-310.

Optimisation of conceptual catchment models
T. O'Donnell 1966-70

Development of techniques of automatic parameter adjustment when fitting conceptual computer models to specific catchments.

Ibbitt, R. P., O'Donnell, T. 1971. Fitting methods for conceptual catchment models. *Am. Soc. civ. Engrs., 97,* HY9 *8370,* 1331-1342.

Ibbitt, R. P., O'Donnell, T. 1974. Designing conceptual catchment models for automatic fitting methods. *Mathematical models in Hydrology 2, Int. Ass. scient. Hydrol, 101, Proc. of Int. Symp. Warsaw 1971.*

Ibbitt, R. P. 1972. Effects of random data errors on the parameter values for a conceptual model. *Wat. resour. Res. Wash., 8,* 70-78.

Design of motorway drainage systems
T. O'Donnell, Dr M. J. Hall
(Dr C. J. Swinnerton) 1968-71

Two methods of forecasting runoff hydrographs from lengths of roadway in cuttings were prepared using data from 77 storms recorded at 8 sites on the M1, M6 and M45 motorways. The first method, a dimensional hydrograph technique, may be used with design storms of the type which generate a single-peaked hydrograph. The second method, a conceptual modelling technique, is more appropriate for complex rainfall patterns which give rise to multi-peaked hydrographs.

Swinnerton, C. J., Hall, M. J., O'Donnell, T. 1972. A dimensionless hydrograph design method for motorway stormwater drainage system. *J. Instn. highway Engrs.. 19, 11,* 2-10.

Swinnerton C. J., Hall, M. J., O'Donnell, T. 1973. Conceptual model design for motorway stormwater drainage. *Civ. Engrg. pub. wks. Rev., 68,* 124-132.

Swinnerton, C. J., Hall, M. J., O'Donnell, T. 1973. Runoff estimation for motorway carriageways running in cuttings. *Seminar on Detailed Road Design, P. T. R. C. Summer Annual Meeting, Brighton, 25-29 June,* 12pp.

Shot noise processes as stochastic models of daily streamflow data
T. O'Donnell, P. E. O'Connell, with Prof. D. R. Cox,
Dept. of Mathematics (G. Weiss) 1970-73

A new class of stochastic models, called shot-noise models, has been proposed for generating daily streamflow data. Particular advantages of shot noise models are their ability to reproduce recession and base flow effects and to generate non-negative flows without invoking transformations. Multisite shot noise models have been developed, and a method for handling non-positive definite sample correlation matrices has been derived.

Weiss, G. (In press). Shot noise models for synthetic generation of multisite daily streamflow data. *Proc. int. Symp. Design of Water Resources Projects with Inadequate Data, Madrid,* 1973 UNESCO/WMO.

Analysis of hydrological time series
T. O'Donnell, Dr M. J. Hall, P. E. O'Connell 1968-73

The derivation of stochastic models for mean daily river flow data was investigated for the simulation of complex water resources systems, and the construction of flood magnitude-frequency relationships. The regionalisation of mean monthly river flow data and its use in feasibility studies for water supply projects was also studied.

Hall, M. J., O'Connell, P. E. 1972. Time-series analysis of mean daily river flows. *Wat. & wat. Engrg., 76,* 125-133.

O'Donnell, T., Hall, M. J., O'Connell, P. E. 1972. Some applications of stochastic hydrological models. *Proc. int. Symp. on Math. Modelling Techniques in Wat. Res. Systems, Ottawa,* 227-239.

Analytical model for transient floods in the presence of persistent baseflow
F. V. Appleby 1965-

An explicit non-linear analytical model is being constructed for total stream flow runoff in terms of jointly acting sub-systems of persistent and transient components with physically significant parameters (time variant or not) as behavioural indicators in the mean for varying catchment conditions and types.

Appleby, F. V. 1970. Recession and the baseflow problem. *Wat. resour. Res. Wash., 6, 5,* 1398-1403.

Appleby, F. V. 1974. A determinate model for runoff on a non-linear system. *Mathematical Models in Hydrology, 2. Int. Ass. scient. Hydrol. 101 Proc. int. Symp. Warsaw, 1971.*

Hydrological consequences of urbanisation
Dr M. J. Hall (B. J. Greenfield, D. P. Carr, J. M. Bleek, A. E. Jones) 1969-74

The finite-period unit hydrograph was used as an index of the changes in flow regime of catchment areas with urban development. A forecasting technique based upon a dimensionless hydrograph with lag time as the only scaling parameter was devised to assist in the design of flood alleviation works for developing urban areas.

Hall, M. J. 1973. Synthetic unit hydrograph technique for the design of flood allevation works in urban areas. *Proc. int. Symp. on Design of Wat. Resour. Projects with Inadequate Data, Madrid, 1,* 145-161.

Hall, M. J. (In press). The hydrological consequences of urbanisation. *Proc. C.I.R.I.A. Colloquium on Rainfall, Runoff and Surface Water Drainage of Urban Catchments, Bristol.*

Velocity fluctuations in natural stream channels
Dr M. J. Hall (P. M. Johnstone, U. H. Mudallal, V. C. Slator, R. G. Hocker) 1970-74

Prolonged series of point-velocity measurements were used to investigate the structure of the low-frequency turbulent fluctuations which give rise to errors in velocity-area methods of stream gauging. Measurements were made on two rivers of widely different character and with both propeller and cup-type current meters. Analysis of the time series of point velocities obtained indicate a concentration of variance at low frequencies but no obvious periodic components.

Hall, M. J., Johnston, P. M. (In press). Stochastic analysis of velocity fluctuations in a natural stream channel. *Proc. int. Symp. Math. Models in Hydrol., Warsaw, 1971*.

Drainage of peat lands
Dr M. J. Hall, with Dr T. M. Prus-Chacinski,
C. H. Dobbie and Partners

Study of records of rainfall and pumping volumes for the Crossens area, a low-lying agricultural catchment of which approximately 40% is covered by peat, has demonstrated the possibilities for incorporating variations in antecedent catchment wetness into simple forecasting procedures for flood volumes. The results obtained may prove to be of more general use for peatlands under similar conditions of climate and agricultural practices.

Hydrology of laboratory catchments
Dr M. J. Hall (formerly T. O'Donnell)
(P. M. Johnston (formerly Dr I. Cordery)) 1968-75

A rainfall simulator with a working area of approximately 11 × 7m has been constructed with the object of studying rainfall-runoff relationships using small artificial catchment areas in the laboratory. The apparatus can be programmed to produce a wide variety of rainfall patterns, and all experimental data are recorded directly in computer-compatible form.

Hydrology of channel networks
Dr M. J. Hall (formerly T. O'Donnell)
(P. M. Johnston) 1971-74

The relationship between the geomorphology of channel networks and the characteristics of the outflow hydrograph is important in the estimation of the flow regime in ungauged catchments. An intensive study of the significant effect on such relationships of different ways of measuring and identifying the geomorphological parameters is being made. Computer simulation experiments show which parameters of network geometry are the most useful in the prediction of hydrograph characteristics.

Model selection and performance for generating synthetic hydrological data
P. E. O'Connell, with Prof. D. R. Cox,
Dept. of Mathematics (P. Prescott) 1973-76

An algorithm is being produced to guide the selection of a suitable model for generating synthetic hydrological data and to assess the success in fitting the model. The responses of direct supply, pumped storage and multi-purpose reservoirs to synthetic data inputs from several models are being investigated. The choice of a suitable time unit for simulation is being considered.

Small sample properties of stationary stochastic models in hydrology
P. E. O'Connell (Z. Sen) 1972-74

The small sample properties of a number of stochastic models of hydrological processes have been derived analytically. Included in these properties are the variance, standard deviation, lag-one autocorrelation coefficient and adjusted and rescaled ranges. Small sample estimates of these quantities from annual streamflow sequences can be corrected for bias using results from this study.

Stochastic models of monthly streamflow

P. E. O'Connell (D. L. Chambers, P. S. Lee) 1969-

Stochastic techniques for simulating monthly streamflow at one or several gauging sites have been investigated. Particular emphasis has been placed on the multisite case, and one well known approach, with an associated computer package, has been found to have a fundamental error incorporated. A multisite extension of the Thomas-Fiering model has been developed to cope with the case where the records of monthly flows at several sites form a monotone sample.

O'Connell, P. E. 1973. Multivariate synthetic hydrology: a correction. *Proc. Am. Soc. civ. Engrs., J. hydraul. Div. 99* (*HY12*), 2393-2396.

Decision making in the design of water resource systems

P. E. O'Connell 1973-

The choice of a stochastic model for generating synthetic streamflows is frequently based on conventional statistical goodness-of-fit criteria. However, such an approach does not necessarily lead to optimal decisions by the water resources engineer. Decision theory affords a means of making an optimal choice of model.

O'Connell, P. E., Wallis, J. R. (In press). Choice of generating mechanism in hydrology with inadequate data. *Proc. int. Symp. Design of wat. Resour. projects with inadequate data, Madrid, 1973 UNESCO/WMO.*

Stochastic modelling of long-term persistence in streamflow sequences

P. E. O'Connell 1971-74

A simple stochastic model has been proposed which is capable of reproducing the observed short-term behaviour of annual streamflow. The model, a mixed autoregressive-moving-average process, has been specifically developed for generating synthetic flows for use in the design and operation of water resource systems. The formulation of the model for generating flows at several sites has received particular attention, as have difficulties arising in estimating the parameters of the model.

O'Connell, P. E. (In press). A simple stochastic modelling of Hurst's law. *Mathematical Models in Hydrology, 1,* 169-187. *Int. Ass. scient. Hydrol, 100, Proc. int. Symp. Warsaw, 1971.*

O'Connell, P. E. (In press). Simulation techniques in stochastic hydrology. *Tran. Inst. Brit. Geogr.*

Conceptual models of the rainfall-runoff process

P. E. O'Connell, with Prof. J. E. Nash,
University of Galway 1968-70

An attempt was made to model the rainfall-runoff process for the Brosna catchment at Ferbane Ireland. Optimum values of model parameters were obtained using an automatic parameter adjustment technique. Different models were compared and it was found possible to account for about 80% of the initial variance of the discharge rate by very simple models.

O'Connell, P. E. Nash, J. E , Farrell, J. P. 1970. River flow forecasting through conceptual models. Part II – the Brosna Catchment at Ferbane. *J. Hydrol., 10,* 317-329.

Areal rainfall from mathematically-defined surfaces
Miss E. M. Shaw (P. P. Lynn, P. S. Lee, A. E. Jones) 1972-

The determination of areal rainfall for time periods ranging from a year down to 2 minutes for natural catchments and for storm limits has been made by the integration of the rainfall volume under surfaces defined mathematically from a network of point measurements. The method using multiquadrics was found to be most suitable for irregularly spaced data and while the hyperboloids of two sheets give the most satisfactory results, the right circular cones are the most efficient. The technique has been extended to automated DAD analysis.

Shaw, E. M., Lynn, P. P. 1972. Areal rainfall evaluations using two surface fitting techniques. *Hydrol. Sci. Bull., 17,* (4). 419-433.

Lee, P. S., Lynn, P. P., Shaw, E. M. 1974. Comparison of multiquadric surfaces for the estimation of areal rainfall. *Hydrol. Sci. Bull., 18.* (3).

Rainfall storm models in hydrology
Miss E. M. Shaw 1972-

The use of rainfall storm models, ranging from dynamically-based to empirical and probablistic, in hydrological forecasting and prediction is being investigated. Use is made of meteorological data to provide design storms (in terms of DAD values) and short-term forecasts.

Department of Geology (ICLGL)

Groundwater studies in East Anglia
M. H. de Freitas, with members of the Water Industry (P. J. Chilton, R. R. Harris, Dr R. Depledge, G. M. Doughty) 1971-73

Hydrogeological and groundwater resource studies have been made of the Upper Yare and Dove Catchments, also of a well at Melton, Cromer, Norfolk. A chemical survey has been made of groundwater of the Glaven catchment.

Underground disposal of waste with special reference to the UK
M. H. de Freitas (Miss M. McKinlay) 1973-76

Study of the hydrogeological problems associated with the disposal of fluid waste underground in the UK.

Influence of reservoirs on groundwater flow
Prof. J. L. Knill 1964-

Collation of data on the influence of reservoir impounding on groundwater conditions, based on reservoirs in UK and Iran.

Knill, J. L. 1971. Environmental, economic and engineering factors in the selection of reservoir sites, with particular reference to Northern England. *Yorks. geol. Soc.*, 124-143.

Knill, J. L. 1972. Assessment of reservoir feasibility. *Quart. J. engng. Geol., 4,* 355-365.

Knill, J. L. 1972. Engineering geology in reservoir construction in Southwest England. *Proc. Ussher Soc., 2.* (5), 359-371.

KING'S COLLEGE

Department of Civil Engineering (KCLCE)

Modelling of hydrologic systems by the finite element method

Dr J. K. White (A. W. Jayawardena) 1973-75

The finite element method is applied to the response to rainfall of surface drainage and groundwater flows by a model both distributed and deterministic with parameters identifiable as properties of the catchment. The model is intended to have an extensive predictive capacity of value in the design and control of water resource systems.

White, J. K. 1973. A finite element deterministic catchment model. *Proc. int. Conf. Numerical Methods in Fluid Mechanics. Southampton. 1973.*

Department of Geography (KCLGG)

Mass movement on the Dorset coast

Dr D. Brunsden, with D. K. C Jones (London School of Economics) 1966-75

Study of rainfall, soil moisture and pore-water pressure has been included in this assessment of a budget and time scale for mass movement activity.

Brunsden, D., Jones, D. K. C. 1972. The morphology of degraded landslide slopes in S.W. Dorset. *Quart. J. engng. Geol., 5.* (3), 205-222.

Brunsden, D. 1974. A systems approach to the study of landslides. *Proc. I.R.P.I. Conf. on Slope Stability and Conservation,* Naples-Cosenza, Italy. 1971.

Brunsden, D. 1974. The degradation of a coastal slope in Dorset, UK. *Inst. Brit. Geogr. spec. Pub., 7.*

QUEEN MARY COLLEGE

Department of Geography (QMCGG)

Environmental controls on stream discharge characteristics: an example from the Waller's Haven, Sussex

C. Everard (J. B. Chatterton, now MIPGG) 1969-73

Contemporary data of three contributary catchments were collected to analyse hydrological response, particularly during storm recessions, under different land use and antecedent moisture conditions. Historical discharge hydrographs were separated into flow components to assess the effects of deforestation and afforestation on discharge characteristics.

Meso-scale investigation of frontal precipitation

Dr B. W. Atkinson, with Dr P. A. Smithson, SHFGG 1969-74

Meso-scale circulations and precipitation distributions in warm fronts are studied by mapping of short-period precipitation amounts using autographic data from about 350 gauges in England and Wales, isentropic analysis, cross-section and tephigram analysis and by application of Canadian work on cloud dynamics to evaluation of meso-scale systems.

Atkinson, B. W., Smithson, P. A. 1972. An investigation into meso-scale precipitation distributions in a warm sector depression. *Quart. J. roy. met. Soc., 98*, 353-368.

Atkinson, B. W., Smithson, P. A. 1974. Meso-scale circulations and rainfall patterns in an occluding depression. *Quart. J. roy. met. Soc., 100*, 3-22.

Atkinson, B. W., Smithson, P. A. 1974. Rain pulses in a warm sector. *Weather, 29*, 44-53.

UNIVERSITY COLLEGE

Department of Geography (UCLGG)

Hydrological impact of urbanisation
Prof. E. H. Brown (G. E. Hollis) 1969-73

The impact of the construction of Harlow New Town on the water yield, flow regimes, flood frequency magnitudes and channel changes of Canon's Brook and on reservoir sedimentation was studied.

Hollis, G. E. 1974. The effects of urbanization on floods in the Canon's Brook, Harlow, Essex. *Inst. Brit. Geogr. spec. Pub., 6*, 123-139.

Hollis, G. E. 1974. River management and urban flooding. In: *Conservation in Practice*, Ed. A. Warren, B. Goldsmith. Wiley. 201-215.

Quantitative analysis of soil erodibility based on physical and chemical measurements
Dr A. Warren, Dr J. Catt, Dr A. Weir
(E. Badokway-Faith)

Rainfall infiltrometer studies are included in this investigation of soil cementing agents.

Department of Geology (UCLGL)

Hydrogeology of the Triassic strata of the Wirral Peninsula
G. P. Jones (B. A. Memon) 1970-

Hydrogeology of Jurassic aquifers
G. P. Jones (R. H. Al-Dabbagh,
M. J. O'Shea, R. N. Reed) 1971-

Hydrogeological investigations are being made in the following areas: Cotswolds around Cirencester (R. H. Al-Dabbagh); in parts of Dorset and Somerset (M. J. O'Shea); the Windrush and adjacent catchments in the Cotswolds (R. N. Reed).

Mathematical modelling in groundwater studies
G. P. Jones, Y. T. Maini (S. Al-Assaf) 1972

Distribution of saline groundwaters in S.E. England
G. P. Jones (K. Chantara)

MANCHESTER

Department of Engineering (MANCE)

Evaporation estimation in a stratified atmosphere
Dr H. S. Takhar (M. W. Liddament,
1971-74) 1971-79

Evaporation from open water surfaces in non-neutral atmospheric conditions is theoretically estimated. An experimental verification of the theory is attempted.

Department of Geography (MANGG)

Role of an upland peat bog as a streamflow regulator in the hydrological system of a small catchment

Dr R. H. Johnson (I. Dyson) 1973-75

By measuring precipitation and stream flow in a small catchment in Langendale, it is hoped to determine the exact role of the peat cover as a regulator of stream flow in a small Pennine drainage system. A network of closely-spaced rain gauges and a stream discharge station have been installed and measurements will be taken for at least a year.

MANCHESTER TECHNOLOGY

Department of Civil and Structural Engineering (MNICE)

Probability aspects of reservoir operations

J. B. White (P. A. McCullen) 1972-75

Transition probability matrix methods of analysis are to be extended to cases of reservoir storage in which input or output depends on content. The effects of degrees and levels of rationing are being investigated also problems of priority for user and evaporation losses. The method appears to have application to other problems of reservoir management. Some parallel work on the use of synthetic flow series is being undertaken.

Hydrogeology of Permian and Triassic aquifers in Northern England

Dr F. T. Howell, with Dr. R. D. Sedgwick, Dept. of Chemistry (Dr C. J. Bow, Dr C. J. Payne, Dr P. J. Thompson, Dr J. M. Crook, Dr F. A. Woodhead, P. Howsam, R. W. Ewers) 1965-

Evaluation of the mechanisms and magnitudes of flow, storage and distribution of fresh and saline waters.

Bow, C. J., Howell, F. T., Thompson, P. J. 1970. Permeability and porosity of unfissured samples of Bunter and Keuper Sandstones of South Lancashire and North Cheshire. *Wat. & wat. Engng* 464-467.

Crook, J. M., Howell, F. T. 1970. Three new simple tests for measuring and estimating the permeability of the Permo-Triassic Sandstone of Northwest England. *Geotechnique*, 446-451.

Crook, J. M., Howell, F. T. 1970. The characteristics and structure of the Permo-Triassic Sandstone aquifer of the Liverpool and Manchester industrial region of Northwest England. *Int. Symp, Groundwater, Palermo*, 217-225.

Crook, J. M., Daw, G. P., Howell, F. T., Morgan, F. R. 1971. Permeation properties of unfissured Bunter Sandstone of Lancashire and Yorkshire. *Geotechnique*, 256-259.

Crook, J. M., Howell, F. T. 1971. Field sampling for engineering tests on Bunter Sandstones. *Ground Engng.*

Crook, J. M., Howell, F. T. 1971. Possible districts in South Lancashire and North Cheshire where the Permo-Triassic Sandstone aquifer appears capable of further development. *Wat. & wat. Engng*, 470-471.

Bow, C. J. Howell, F. T., Payne, C. J., Thompson, P. J. 1972. A permeameter for investigating the passage of water through unfissured samples of Permo-Triassic sandstone. *Civ. Engng.*, 261-263.

Howell, F. T., Woodhead, F. A. 1972. A null method for the estimation of the permeability of irregular specimens of permeable strata. *Geotechnique*, 352-356.

Howell, F. T. 1973. The sub-drift surface of the Mersey and Weaver Catchment. *Geol. J., 8,* 2, 285-296.

Crook, J. M. Howell, F. T., Woodhead, F. A., Worthington, P. F. 1973. Permeation properties of Bunter Sandstones from the Cheshire and Fylde Basins. *Wat. & wat. Engng.* 220-221.

Crook, J. M., Howell, F. T., Woodhead, F. A. (In press). The effect of applied stress upon the permeability of some Permian and Triassic sandstones of Northern England. *Third int. Cong. Rock Mechanics, Denver,* 1974.

Crook, J. M., Howell, F. T. (In press). Pleistocene glaciers as the source of fossil freshwaters encountered in the Permian and Triassic Sandstones of the Cheshire Basin, England. *Int. Symp. Development of Groundwater Resour. Madras, India,* 1973.

Department of Physics (MNIPH)
Microphysics of rainfall
Prof. J. Latham, Dr C. P. R. Saunders
(Dr S. G. Jennings, Dr P. R. Brazier-Smith, Dr A. G. Crowther, Dr M. H. Smith, Dr R. F. Griffiths, N. A. M. Wahab, M. J. Gay, R. L. Reed, B. S. Wong) 1968-

Experimental and theoretical studies are being made of the microphysics of rainfall. Problems studied: coalescence efficiencies of raindrops; production of satellite drops by colliding raindrops; growth and dissipation of a rain-shaft; aggregation of ice crystals, and vibration and fall velocities of raindrops. Other research includes: the construction of a novel form of fast-response raindrop distrometer and field studies of orographic influences on rainfall from a Pennine field station.

Latham, J., Myers, V. 1970. Loss of charge and mass from raindrops falling in intense electrical fields. *J. geophys. Res., 75,* 515-520.

Latham, J., Saunders, C. P. R. 1970. Experimental measurements of the collection efficiencies of ice crystals in electric fields. *Quart. J. roy. met. Soc., 96,* 257-265.

Smith, M. H., Griffiths, R. F., Latham, J. 1971. The freezing of raindrops falling through strong electric fields. *Quart. J. roy. met. Soc., 97,* 495-505.

Latham, J. 1971. A quantitative assessment of precipitation mechanisms of thunderstorm electrification. *J. met. Soc. Japan, 48,* 359-366.

Brazier-Smith, P. R., Jennings S. G., Latham, J. 1971. Accelerated rates of rainfall. *Nature Lond., 232,* 112-113.

Brazier-Smith, P. R., Jennings, S. G., Latham, J. 1972. The interaction of falling water drops: Coalescence. *Proc. roy. Soc. Lond., A326,* 393-408.

Brazier-Smith, P. R., Jennings, S. G., Latham, J. 1973. Raindrop interactions and rainfall rates within clouds. *Quart. J. roy. met. Soc., 99,* 260-272.

Brazier-Smith, P. R., Jennings, S. G., Latham, J. 1973. The influence of evaporation and drop-interactions on a rainshaft. *Quart. J. roy. met. Soc., 99,* 704-722.

Crowther, A. G., Saunders, C. P. R. 1973. On the aggregation and fragmentation of freely falling ice crystals in electric fields. *J. met. Soc. Japan, 51,* 490-493.

Chisnell, R. F., Latham, J. 1974. A stochastic model of ice particle multiplication by drop splintering. *Quart. J. roy. met. Soc., 100,* 296-308.

MIDDLESEX POLYTECHNIC

Department of Geography and Planning (MIPGG)

Flood hazard research project
Dr. E. C. Penning-Rowsell, D. J. Parker (Dr J. B. Chatterton, Miss S. Davies, S. Farrell) 1970-76

The damage resulting from flooding is being investigated, as an input to evaluations of flood alleviation measures to secure maximum use of flood plain areas, by land use survey, flood damage potential surveys for different land uses, correlations with published data on land use and correlations with flood extent and frequency data.

Penning-Rowsell, E. C., Parker, D. J. 1974. Improving flood plain development control. *J. roy. town Planning Inst.*, *60*, 2, 540-544.

NATIONAL COLLEGE OF AGRICULTURAL ENGINEERING, SILSOE

Department of Field Engineering (NCAFE)

Soil erosion in U.K.
Dr R. P. C. Morgan

Field studies of rainsplash, overland flow and rill wash in the Silsoe area form part of this research into soil erosion.

Morgan, R. P. C. 1973. A field study of soil erosion near Silsoe, Bedfordshire. *Basic Asset*, *1*, 17-18.

Morgan, R. P. C. 1974. Nature provides, man erodes. *Geogr. Mag.*, *46*, 528-35.

Prediction of runoff from small catchments in Britain
W. B. J. Withers (R. D. L. Beck) 1971-73

Techniques for prediction of peak flow from small rural catchments in Britain for purposes of design of engineering works (e.g. culverts, drainage channels) were evaluated and developed. Data from a small clay catchment were used for a comparative test of four prediction methods.

NEWCASTLE-UPON-TYNE

Department of Civil Engineering (NEWCE)

Friction coefficients for flows with small depth and high roughness
Prof. P. Novak 1972-

Friction coefficients in open channel flows with large roughness and small depth have been studied to provide a rational approach to roughness size and depth evaluation and some insight into the turbulence characteristics. A relationship between the constants of the friction coefficient equation and the concentration and shape of the roughness elements has been obtained, also information about the plane of reference for the channel bed. Different concentrations and configurations of spherical particles are being used as roughness elements.

Channel morphology and runoff
Prof. P. Novak (P. Adamson) 1972-73

Stream systems, channel morphology and runoff have been studied and the variability of river flow, expressed in terms of its standard deviation, variance, serial dependence and 6-hour flow duration curve, was found to be related to the discharge per unit area of the catchment and the density and spatial distribution of the drainage network. The distribution of rapid water level changes within the annual hydrograph was shown to be related to catchment area or mean flow and study is continuing. A pilot study of the effect of flood hydrograph shape on the channel morphology was carried out in a trapezoidal flume filled with sand.

Pits for bed-load measurement
Prof. P. Novak 1972-73

A comprehensive survey of bedload measurement techniques and a laboratory study aiming at design parameters of sediment pits (traps) to be used on smaller streams in an upland catchment have been completed.

Groundwater movement resulting from mineral abstraction
R. E. Featherstone (A. M. Al-Sammarae) 1972-75

Hydrogeological characteristics of a confined sulphur-bearing stratum are being studied both in the undisturbed condition and after sulphur abstraction. From field data, natural conditions have been simulated using an electrical resistance analogue and the simulation of groundwater flow and changing permeability due to abstraction is being investigated.

Unit response functions
P. Johnson (M. Clements, D. Reed) 1970-

Efficient identification of unit response functions (e.g. unit hydrographs) is being investigated by techniques based on Laplace transforms and on least squares polynomials. Further work involves study of the variation of unit response functions between storm events and methods whereby this variation can be incorporated initially into identification techniques and then into flood forecasting.

Johnson, P. 1970. Calculation of the instantaneous unit hydrograph using Laplace transforms. *Proc. int. Symp. on Results of Research on Representative and Experimental Basins, Wellington, J. Hydrol.* (*New Zealand*).

Johnson, P., Clements, M. 1971. Determination of instantaneous unit hydrographs by least squares polynomials. *Int. Symp. on Mathematical Models in Hydrology, Warsaw.*

Reed, D. (In press). A non-linear rainfall-runoff model, providing for variable time lag. *J. Hydrol.*

Hydrochemistry and its potential use in hydrology
P. Johnson (A. Davies, B. Glover) 1969-74

Optimisation methods are used universally in determining suitable values of catchment model parameters. The possibility of using variations of naturally-occurring chemical concentrations in river water to identify streamflow components and thereby to determine catchment model parameters is being studied.

Glover, B., Johnson, P. (In press). Variations in the natural chemical concentrations of river water during flood flows and the lag effect. *J. Hydrol.*

Simulation by conceptual modelling
P. Johnson (Dr M. Thoms, H. Suyubatmaz, Miss S. Slator) 1966-

A model of the precipitation runoff process of natural catchments on a digital computer is giving acceptable accuracy of simulation whilst restricting the number of model parameters used. So far the number of parameters used has not exceeded six (including one for the snowmelt process). This water-balance model is generally used on a daily data basis and now simulates in a spatially-distributed mode. Plans are made to integrate this model with dynamic modelling of catchment response.

Snowmelt river flooding in the British Isles
P. Johnson, D. Archer 1971-74

The significance of snowmelt to British hydrology has been investigated and a technique developed whereby forecasts of river floods resulting from major snowmelt events, with or without incident rainfall, can be made. The well known degree-index method of prediction has been studied and variation of the index coefficient over more than 20 catchments in different parts of Britain has been correlated with snowmelt volume, catchment area and slope. A conceptual model for distributed energy inputs, spatial variation of snow on the catchment and for variable time lag in runoff response to melt has been developed to improve forecasting.

Johnson, P., Archer, D. 1972. The hydrological significance of snow in Britain. *Proc. int. Symp. on Role of Snow and Ice in Hydrology, Banff.*

Johnson, P., Archer, D. 1972. Current research in British snowmelt river flooding. *Hydrol. Sci. Bull., 17*, 4, 12.

Johnson, P., Archer, D. 1975. Snowmelt runoff. *Flood Studies Report*, Chap. 7. Natural Environment Research Council, London.

Johnson, P., Archer, D. (In press). *Snow hydrology and river flooding in Britain.* Univ. Newcastle-upon-Tyne.

Stochastic-deterministic models
P. Johnson (Dr B. M. J. Barton, P. E. Jones) 1967-

The concept of generating monthly hydrometeorological data and routing them through a deterministic model of a catchment's water balance to synthesise monthly river flow has been studied. Data from over 180 catchments in Britain were used to develop the technique. Stochastic parameters were regionalised and deterministic parameters were correlated with catchment topography, geology and soil type. Similar mixing of stochastic-deterministic processes is being considered on a daily and spatially distributed basis.

Barton, B. M. J. 1973. Synthetic monthly runoff records for ungauged British catchments. *J. Instn. wat. Engrs.*

Groundwater resources
P. Johnson (J. C. Eastwood, M. Briggs) 1970-74

General water resource potential of two distinct groundwater environments has been studied. In both cases, simple deterministic water balance models have been developed to simulate rainfall-runoff with the replenishment of groundwater forming an important feature of the overall synthesis. The first groundwater basin is that of the chalks underlying the river Stour in Dorset. The second considers the vast volumes of water held in the old coal workings of the Northumberland and Durham coal fields.

Department of Geography (NEWGG)

Crop and soil water relationships in an agricultural catchment, Northumberland.
Dr L. W. Hanna (N. Siam) 1973-78

Meteorological recordings in the catchment allow estimation of daily Penman Eo and rainfall. Soil water is measured by neutron probe calibrated by field sampling. Runoff is monitored by stream gauging. Several crops (currently Italian rye grass and winter wheat) will be studied for water use and an attempt will be made to assess the influence of crop type on the hydrology of the catchment. The results will be used for a more general land use model for water resource management.

Effective rainfall reaching a stream channel in an upland catchment
Dr H. Lister (J. Catterall) 1972-76

Measurements are made of climatic data including thermal radiation and radiation flux at ground level; vertical profiles of humidity, temperature and wind speed for short term evaporation flux, soil moisture content, soil structure and stream discharge for heat and water budgets. Sub-surface flow, although localised along reaches of stream channel, is not found in discrete 'pipes' and so is estimated not measured.

NOTTINGHAM

Department of Civil Engineering (NOTCE)

River flood frequencies
Dr F. Rieband, with M. S. T. Chang,
Trent River Authority 1966-69

A flood frequency analysis has been carried out with particular reference to geological grouping of catchments. A method for estimating the effect of flood frequencies of catchment slopes, based on 'critical durations', has been developed. Digital computing has been used and the results of the geologically-grouped flood frequency analysis are presented in design chart form.

School of Agriculture

Department of Physiology and Environmental Studies (NOTSA)

Evaporation from a catchment
Prof. J. L. Monteith, Dr M. McGowan,
Dr. M. H. Unsworth (Dr N. A. B. Trivett,
Dr J. Williams, G. Russell) 1968-74

Estimates of evaporation from a catchment (Kingston Brook) were compared with rainfall and runoff measured by the Trent River Authority. The use of water by different types of vegetation was explored and the depth of water extraction by roots related to plant and soil characteristics. The effects of atmospheric pollution on evaporation rates through the radiation balance were assessed. Soil water content was measured with a neutron probe and a method was developed to estimate drainage and evaporation rates separately. Soil water potential was measured with tensiometers and thermocouple psychrometers. Diurnal and seasonal changes of surface resistance have been calculated for grass, barley and woodland.

Unsworth, M. H., Monteith, J. L. 1972. Aerosol and solar radiation in Britain. *Quart. J. roy. met. Soc., 98*, 778.

McGowan, M. 1974. Depths of water extraction by roots: application to soil-water balance studies. In: *Isotope and radiation techniques in soil physics and irrigation studies*, International Atomic Energy Agency, Vienna, p. 435.

OXFORD

Department of Agricultural Science (OXFAS)

Effect of field drainage and climatic change on water levels

F. H. W. Green 1970-

Estimation of rainfall amounts, within practically useful limits, for about four months in advance, for early summer, and for mid-winter, by the use of correlations which appear to have been effective in north-west Europe since the nineteen-forties.

Green, F. H. W. 1972. Characteristics of precipitation in the Scottish Highlands. *Distribution of precipitation in mountainous areas, II. Geilo Symp. Norway. WMO/OMM*, 140-146.

Green, F. H. W. 1973. Recent summers in the Scottish Highlands. *Weather, 28*, 430-434.

Green, F. H. W. 1973. Aspects of the changing environment. Some factors affecting the aquatic environment in recent years. *J. env. Managmt., 1*, 377-391.

Effect of application of fertilizers on soil, well and stream water

F. H. W. Green 1970-

Green, F. H. W. (In press). Changes in artificial drainage, fertilizers and climate in Scotland. *J. env. Managmt. 2*.

Study of changes in the incidence of snow in Britain and southern Norway

F. H. W. Green 1970-

Green, F. H. W. 1970. Seasonal changes of snow cover in the Cairngorms. *Weather, 25*, 211-213.

Green, F. H. W. 1973. Changing incidence of snow in the Scottish Highlands. *Weather, 28*, 386-394.

Potential evaporation measurement

F. H. W. Green 1970-

Management of network of lysimeter observations in Britain for determining potential evaporation and potential water deficit/surplus.

Green, F. H. W. (Annually). Potential evapotranspiration measurements. *Brit. Rainfall.*

Green, F. H. W. 1973. Raingauge accuracy – the evidence from lysimeter observations. *Weather, 28*, 386-394.

Department of Forestry (OXFFY)

Water relations of forest trees

Dr L. Leyton, F. B. Thompson 1964-

The heat exchange method of measuring sap flow in trunks to monitor transpiration from trees is being further developed to remove errors associated with changes in heat storage. Measurements will be made of internal water stress and stomatal regulation.

Leyton, L. 1970. Problems and techniques in measuring transpiration from tree crops. In: *Physiology of Tree Crops*. (Eds: L. C. Lackwill and C. V. Cutting). Academic Press, Canada. 101-111.

Thompson, F. B., Leyton, L. 1971. Method for measuring the leaf surface area of complex shoots. *Nature, Lond.* 229-572.

Hydrological relations of forest and moorland vegetation

F. B. Thompson 1965-76

Work on moorland vegetation, involving measurement of cover density, interception, storage capacity and duration of wetness is being completed. The effect of the deciduous habit in trees and the seasonal climate on rainfall interception is being investigated.

Thompson, F. B. 1972. Rainfall interception by oak coppice (*Quercus robur L*). In: *Research papers in forest meteorology*. (Ed: J. A. Taylor), Cambrian News (Aberystwyth) Ltd. 59-74.

Hydrological interpretation of plant root distributions

Dr E. R. C. Reynolds (Miss M. Brown) 1964-

Employing pattern analysis, the variations of rooting along transects and down profiles have been examined in a plantation of Douglas fir. In the same plantation, an underground glazed observation chamber has been constructed to assess the rate and variability (both in time and space) of root growth and thus the mode of exploitation of the soil moisture reserves.

Reynolds, E. R. C. 1970. Evaluating alleged external physical benefits of forests. *J. Oxf. Univ. for. Soc. (7th Ser.), 1,* 17-20.

Reynolds, E. R. C. (In press). The distribution pattern of the fine roots of trees. *2nd int. Symp., Ecology and Physiology of Root Growth. 1971 Potsdam.*

Reynolds, E. R. C. (In press). Tree rootlets and their distribution. *3rd Cabot Symp., Harvard Forest.*

Department of Geography (OXFGG)

Clear water erosion: its geographical and international significance

Dr R. P. Beckinsale

Beckinsale, R. P. 1972. The effect upon river channels of sudden changes in sediment load. *Acta Geographica. Debrecina, 10,* 181-186.

Beckinsale, R. P. 1972. Clear water erosion: its geographical and international significance. *Int. Geogr., Montreal, 2,* 1244-46. *Univ. of Toronto Press.*

Solution and percolation rates of selected limestones

Dr M. M. Sweeting 1970-

Variations in karst landscape are being studied by petrological examination of different limestones, determination of their effective porosities and solution rates, examination of residues and the karst waters. In Britain,

N.W. Yorkshire is being studied in detail, but studies have been made on dolomites in Ontario, Canada; dolomites in the Transvaal, S. Africa, and some Pleistocene and Pliocene limestones in New Zealand.

Sweeting, M. M. (In press). Lithological variations in limestones and their relationships to karst and caves. *Int. speleolog. Union congress Proc. Czechoslovakia, 1973.*

Studies in hydrology and morphology of selected Cotswold rivers
Dr M. M. Sweeting (M. J. Hart) 1969-74

The rivers Windrush, Evenlode and Coln have been gauged regularly for 2 years and sediment analyses taken. The morphology of the rivers (meanders, straight stretches, bed characteristics) have been examined. Comparisons between rivers on the scarp and dip slopes of the Cotswolds are being made. The R. Leach, which is predominantly a limestone stream, has also had much attention.

PLYMOUTH POLYTECHNIC

Department of Environmental Science (Geography) (PLPES)

Karst water studies in the Central Pennines
Dr J. L. Ternan 1970-

The role of hydrological and hydro-meteorological variables (residence time of ground-water spring discharge, antecedent effective precipitation) in influencing spatial and temporal variations in limestone solution is being investigated in the Malham area north of the Craven faults.

Ternan, J. L. 1972. Comments on the use of a calcium hardness variability index in the study of carbonate aquifers: with reference to the Central Pennines, England. *J. Hydrol., 16,* 317-321.

Ternan, J. L. 1974. Some chemical and physical characteristics of five resurgences on Darnbrook Fell. In: *Limestones and Caves of N.W. England.* (Ed: A. C. Waltham). David and Charles. 115-126.

Halliwell, R. A., Ternan, J. L., Pitty, A. F. 1974. Introduction to the karst hydrology of north-west Yorkshire. In: *Limestones and caves of N.W. England.* (Ed: A. C. Waltham). David and Charles. 106-114.

Integrated study of a small upland granite catchment in S.W. Dartmoor – the Narrator Brook
Dr J. L. Ternan, P. Sims, D. Gilbertson, M. Kent 1973-
A long-term integrated study of the hydrology, microclimatology, geomorphology and ecology of a small upland catchment is being set up with instrumentation which will include 4 autographic raingauge stations, one stream gauging station, and an automatic weather station (Wallingford type) and a soil neutron probe.

READING

Department of Agricultural Botany (RDGAB)

Agricultural consequences of extensive groundwater extraction in E. Anglia.
Dr D. S. H. Drennan 1969-

The effects of groundwater extraction on surface water resources, water-table changes and water in the soil available to crops are being examined in relation to potential damage to crops grown in an area with extensive groundwater development for river regulation.

Department of Geography (RDGGG)

Investigations into throughflow in soils

Dr R. Hardy (Dr B. J. Knapp) 1967-73

Direct investigations of water movement within a soil catena in the Plynlimon catchments were carried out using automatic logging systems. Flow nets were also studied.

Knapp, B. J. 1970. A note on throughflow and overland flow in steep mountain catchments. *Reading Geogr., 1*, 40-43.

Knapp, B. J. 1973. A system for the field measurement of soil water movement. *British geomorpholog. Res. Grp. tech. Bull*, 9.

Knapp, B. J. 1974. Hillslope throughflow observation and the problem of modelling. In: *Inst. Brit. Geogr. Spec. Pub., 6*.

Knapp, B. J. (In press). Infiltration and storage of soil water, Chap. 2. In: *Hillslope Hydrology*. (Ed: M. J. Kirkby). Wiley, London.

Investigation of the sediment and chemical balance of two Welsh catchments

Dr J. R. Hardy (N. C. Oxley) 1969-72

Two streams flowing into Llyn Ebyr, Montgomeryshire, were monitored for discharge and sediment characteristics and chemical analyses of water samples were carried out for estimation of sediment budgets and denudation rates.

Oxley, N. C. 1974. Suspended sediment delivery rates and the solute concentration of stream discharge in two Welsh catchments. In: *Inst. Brit. Geogr. spec. Pub., 6*.

ST ANDREW'S

Department of Geography (STAGG)

Hydrological processes in a Scottish Upland catchment (Glen Devon, Perthshire)

A. Werritty, Dr M. F. Thomas, Dr J. Jarvis (G. W. Rae) 1972-

Hydrologic processes within a small upland peat catchment are being investigated with special emphasis on modes of channel extension, partial area contributions, hillslope/soil hydrology relations, development of gullies, intermittent channels and piping. An automatic meteorological station has been installed together with two gauging stations and a series of automatic throughflow recorders.

SALFORD

Department of Civil Engineering (SALCE)

Simulation of catchment response to rain by mathematical modelling

Prof. E. M. Wilson (S. Zein) 1972-75

Several investigators have produced mathematical models of catchments which reproduce flood discharges under simulated rainfall. The present project is a development of those taking into account the effects of slope and rain intensity on infiltration and runoff. An attempt is being made to provide a general programme which can be quickly adapted to any catchment.

Effect of slope, soil properties and rainfall intensity on infiltration and runoff
Prof. E. M. Wilson (Dr S. Nassif) 1970-74

A 24m^2 weighable model catchment, carrying 5 tonnes soil and capable of being set at various slopes up to 32% was subjected to rain at different intensities, while measurements of infiltration and runoff were made.

SHEFFIELD

Department of Civil and Structural Engineering (SHECE)

Weir calibration by pondage drawdown technique
Dr F. A. Johnson (J. W. East) 1973-76

A laboratory study is being made of the accuracy and range of application of this little used technique. Initial results show that it is ideally suited to the rapid calibration of small flow gauging structures.

Hydrological significance of the natural chemistry of water
Dr F. A. Johnson 1974-77

The inter-relationships between variations in the chemical content of stream waters and the hydrological characteristics of the catchment are being studied, in particular the contribution to flood routing and hydrograph separation techniques.

Flood control in reservoirs and storage pounds
Dr E. J. Sarginson 1972-73

By approximating a natural hydrograph to a single mathematical curve, the routing procedure has been reduced to a dimensionless differential equation. The computer solution, in tabular form, enables the length of spillway necessary to produce any desired reduction in peak flood, or any desired limiting head on the spillway, to be obtained in a single calculation.

Sarginson, E. J. 1973. Flood control in reservoirs and storage pounds. *J. Hydrol., 19,* 351-359.

Statistical treatment of urban drainage
Dr E. J. Sarginson 1972-73

A statistical examination of the data on urban runoff has shown the importance of the catchment slope on the effective impermeability of the area. The analysis has resulted in a version of the rational formula which includes a term for the slope.

Sarginson, E. H. 1973. A statistical treatment of urban drainage design. *J. Instn. municip. Engrs., 100,* 152-154.

Urban rainfall and runoff modelling
Dr E. J. Sarginson, with B. B. Nussey, Sheffield Polytechnic

To find a mathematical model which will relate urban rainfall and runoff, data from this country and USA have been fitted to a model with considerable success but further investigations are needed to eliminate errors.

Sarginson, E. J. 1973. Relationship of rainfall and runoff in urban areas. *CIRIA Colloquium on Rainfall, Runoff and Surface Water Drainage of Urban Catchments, Bristol.*

Department of Geography (SHFGG)
Effects of vegetation on runoff/infiltration ratio
Prof. S. G. Gregory, Dr A. D. Knighton
(A. S. Tricker) 1970-

To assess the relative importance of 4 groups of factors: rainfall characteristics, soil conditions, surface cover, slope, in affecting the infiltration/runoff ratio, measurements are being made of infiltration by cylinder infiltrometer and rainfall simulator in a small basin on interbedded gritstones and shales.

Baseflow component of the streamflow hydrograph
Prof. S. G. Gregory, Dr. A. D. Knighton
(R. Trend) 1970-

Within a third-order basin consisting of interbedded sandstones and shales, the characteristics of, and the factors affecting, baseflow fluctuations were examined.

Generating functions of morphometry in some selected fluvial networks
Prof. R. S. Waters, Dr A. D. Knighton
(P. K. Ray) 1973-75

To identify the morphometric explainability of network growth functions and their probabalistic nature on the phases of network development, data are collected from maps, air photographs and from field surveys. Changes in the networks are traced by comparative study of air photographs of different time series.

The effects of drainage basin morphology on flood characteristics
Prof. R. S. Waters (T. Krasu) 1973-76

Hydrograph forms are being statistically related to basin characteristics.

Solute sources and variations in streams in Mid-Wales
R. Cryer 1970-74

Precipitation and stream water samples are analysed to determine the major ion composition, and solute concentration is related to various hydrological parameters to identify solute sources and movements in a catchment.

Lewin, R., Cryer, R., Harrison, D. I. 1974. Sources for sediments and solutes in Mid-Wales. In: *Inst. Brit. Geogr. spec. Pub., 6*, 73-86.

Mechanics of bank erosion
Dr A. D. Knighton (M. P. Mosley) 1973-

Determination of the main factors affecting the amount and location of bank erosion, particularly where the banks are composed of cohesive material, by measurements on the R. Noe in Derbyshire and the R. Bollin-Dean in Cheshire.

Bed slope adjustment in natural channels
Dr A. D. Knighton 1971-

Investigation of the factors controlling channel slope and the processes of bed slope adjustment, by explaining spatial variation in channel gradient and profile form in terms of selected topographic, sediment and hydraulic variables and by relating processes such as material sorting to bed slope adjustment. Flume studies are involved.

Knighton, A. D. (In press). Form adjustment in a limestone dry valley: *East Midland Geogr., 7.*

Storm rainfall in Northern England, 16 July 1973
Dr P. A. Smithson (see also QMCGG) 1973-

Meteorological reasons for a period of heavy rainfall which gave the Sheffield area its largest recorded daily fall of 119.2mm were investigated by obtaining daily gauge readings and meteorological analysis based on published records.

Smithson, P. A. 1974. Heavy rainfall over Central Northern England. *Weather, 29* (1), 17-24.

Infiltration characteristics of a small Pennine catchment
A. S. Tricker 1970-74

Infiltration characteristics of soil/vegetation complexes within a small drainage basin are studied by collection of field and laboratory data relating to several aspects of the biotic and abiotic environment. Both cylinder and rainfall simulator infiltrometers have been developed and used for the field measurements of infiltration rates.

Department of Geology (SHFGL)

A geochemical investigation of the Claver and Darley Dale areas of east Derbyshire
Dr P. R. Ineson (J. D. Hedges) 1971-74

The influence of past and present mining activities on the trace element content of the soils and waters of the River Derwent and its tributaries is being studied using X-ray fluorescence, atomic adsorption, Dithizone and emission spectrographic analysis.

Influence of the Quaternary deposits on the composition of groundwater in a Triassic aquifer (Vale of York)
Dr D. A. Spears (Mrs S. M. Rhodes) 1972-

Porewater and sediment extracted from boreholes in Quaternary deposits (boulder clays, lacustrine and fluviatile sands and clays) are analysed for major chemical components to establish the source of elements and to compare porewater chemistry with groundwater.

Spears, D. A. (In press). Mineralogy and geochemistry of some Quaternary deposits in the Vale of York. *Proc. Yorks. geol. Soc.*

SOUTHAMPTON

Department of Civil Engineering (SOTCE)

Mathematical models for circulation and dispersion studies in shallow waters
Dr C. A. Brebbia (2 Research assistants, 4 Research students) 1972-

Numerical methods, in particular finite elements, are being used to build computer models for shallow water circulation and dispersion problems.

Brebbia, C. A. 1973. Application of the finite element method to steady irrotational flow. *Revue Roumaine des Sciences Techniques: Ser. Mecan. appliq.,* 3.

Brebbia, C. A. 1973. Finite element solution for effluent dispersion. *Proc. int. Conf. Numerical Methods in Fluid Dynamics, Southampton 1973.*

Brebbia, C. A. 1974. Circulation Problem. *Proc. Finite Element Sem.* Atlas Computer Laboratory, Sci. Res. Council.

Extension of rainfall and runoff records using a multiple regression approach
Dr P. R. Helliwell (Dr R. Bilgin) 1970-74

A multi-site linear data fill-in model was developed to handle data sets of monthly rainfall and runoff values. Catchments studied were the R. Spey (22 rainfall stations), R. Wye (40 rainfall and 11 streamflow stations), and R. Great Ouse (31 rainfall and 17 streamflow stations).

Raingauge performance studies
Dr P. R. Helliwell, with M. J. Green, Water Research Association 1969-74

An experimental programme of work in a large wind tunnel and at a field station in the Thames Valley has been carried out to evaluate errors in point measurement of rainfall. Several conventional gauge profiles were tested. Errors due to disturbance of the airflow by the rain gauge have been studied in detail, and results are in general agreement with those of earlier workers. Adhesion and wetting appear to be causes of significant underestimation of catch.

Green, M. J., Helliwell, P. R. 1972. The effect of wind on the rainfall catch. In: *Distribution of Precipitation in Mountainous Areas,* Geilo Symp., Norway, WMO/OMM 326.

Flow resistence, sediment transport and bed formations in alluvial channels
Dr K. V. H. Smith (Dr M. F. Rifai, Dr C. J. Pratt, Dr T. R. Davies, T. J. Yates) 1966-76

Initially, detailed laboratory observations were carried out on the variation of bed formations in alluvial channels. Together with other laboratory observations and prototype data on a world-wide basis, a procedure for predicting stage-discharge relations in alluvial channels has been developed in a computerised form and is being used with Colby's method for estimating sediment discharge in a study of the factors determining the regime of rivers and canals.

Smith, K. V. H. 1970. Similarity in unlined irrigation canal systems. *Proc. Am. Soc. civ. Engrs.*

Chaudry, M. O., Smith, K. V. H., Vigil, A. 1970. Computation of sediment transport in irrigation canals. *Proc. Instn. civ. Engrs.*

Smith, K. V. H. 1970. The Einstein-Barbarossa diagram for computing bed-form resistance in alluvial channels. *Proc. Instn. civ. Engrs.*

Rifai, M. F., Smith, K. V. H. 1971. Flow over triangular elements simulating dunes. *Proc. Am. Soc. civ. Engrs.,* HY7.

Pratt, C. J., Smith, K. V. H. 1972. Ripple and dune phases in a narrowly graded sand. *Proc. Am. Soc. civ. Engrs.,* HY5.

Smith, K. V. H., Yates, T. J. (In press). Prediction of depth-discharge relation in alluvial channels. *Proc. Instn. civ. Engrs., 2,* 7730.

Analysis of magnitude and variation of irrigation water requirement
Dr Z. Svehlik (M. O. Chaudry, G. L. Morling, D. B. Thapa) 1971

Variations of irrigation water requirement and corresponding hydrological parameters have been investigated by use of a mathematical model based on an irrigation regime simulated by computer.

Department of Geography (SOTGG)
Relationship of precipitation distribution to streamflow
Dr Angela M. Foyster (see EXEGG) 1974-77

An instrumented catchment in the New Forest allows detailed evaluation of both spatial and temporal variations of precipitation, losses and soil moisture content, as well as runoff amount and composition to clarify some of the relationships between precipitation and runoff dynamics both in the study area and, perhaps, more generally.

STRATHCLYDE

Department of Applied Physics (STRAP)

Causes and extent of fog interception in a coniferous forest
Dr T. Woodhead, Dr J. M. Crowther, with Dr J. S. G. McCulloch, Institute of Hydrology (N. M. Hancock) 1974-77

In western areas of Britain, the interception by foliage of fog and mist droplets may make a significant contribution to the water balance. The incidence of fog interception and fog drip in coniferous forest is to be measured and related through a mathematical model to the extent of cloud cover and other meteorological variables, and to tree species, location, slope and other site parameters.

Spatial variability of potential evaporation
Dr T. Woodhead (Miss M. N. Leese) 1970-74

The network density of automatic climate stations to permit a mean catchment potential evaporation rate to be calculated to varying degrees of precision is being investigated. Values of Penman's estimate of evaporation, calculated for 5-day, 10-day and monthly periods in 1968 for 40 meteorological sites in south-east England have been analysed by multivariate analysis to determine predictive correlations between evaporation rate and site indices.

Department of Civil Engineering (STRCE)

Hydraulic geometry of tidal watercourses
Prof. D. I. H. Barr (Dr A. G. Chantler) 1970-73

The geometry of prototype and model estuaries are linked by a mathematical model of estuary formation, which generates the estuary shape by application of regime-type relations to tidal flows initially produced in a basic drainage channel, and in combination with an end effect relationship, which applies to the enlargement of the downstream end of a river flowing into a tideless sea.

Chantler, A. G. (In press). The applicability of regime theory to tidal watercourses. *J. hydraul. Res.*

Mathematical models of the erosion-transport-deposition processes

Dr G. Fleming (M. Fahmy) 1972-75

Conceptual models for the water processes basic to hydrology have been developed and integrated with erosion and deposition models to develop a tool for designing water resource systems and to assess the effect on the hydrological response and sediment yield of changes to natural catchments such as land use, vegetation cover, water resource development.

Fleming, G. 1972. Sediment erosion-transport-deposition: state of the art. *Proc. U.S.D.A. Workshop, Oxford, Mississippi, November 1972.*

Fleming, G. 1973. *The potential of mathematical models to the study of runoff and sediment erosion in Morocco.* Report to UN/FAO., Rome.

Fleming, G., Fahmy, M. 1973. Some mathematical concepts for simulating the water and sediment systems of natural watershed areas. *Report HO -73-26 Dept. of civ. Engn. Strathclyde Univ.*

Study of mathematical relationships relating the soil-water-plant continuum

Dr G. Fleming (Mrs K. Trudgill) 1973-76

A laboratory rainfall simulator is used to study the sediment detachment process as affected by soil size, surface slope, rainfall intensity, soil moisture and vegetation cover. A classification of data requirements for various mathematical models for hydrology and sediment is also included in an attempt to relate data input requirements to the objectives of water resource studies.

Developing and testing mathematical models in hydrology

Dr G. Fleming 1967-

Development, testing and comparison of mathematical models in hydrology including the Stanford, HSP, USDAHL, SSARR, HEC, BOUGHTON, and RRL Models.

Fleming, G. 1970. Simulation of streamflow in Scotland. *Bull. int. Ass. scient. Hydrol., 15, 1,* 53-59.

Fleming, G. 1971. Simulation techniques in vegetation management. *Irrigation and Drainage Conf., Miami, November 1970. Proc. Am. Soc. civ. Engrs., IR2, Paper 8175,* 249-262.

Fleming, G. 1971. Application of hydrologic simulation to water resources planning. XIVth *Int. Ass. hydraul. Res. Cong., Paris, 1971,* 257.

Fleming, G., Franz. D. 1971. Flood frequency estimating techniques for small watersheds. *Proc. Am. Soc. civ. Engrs., HY9, Paper 8383,* 1441-1460.

Fleming, G. 1971. A note of river flow forecasting through conceptual models. *J. Hydrol., 13* (4), 351-356.

Fleming, G. 1973. Real-time operation of a water resource system. *Proc. Int. Ass. hydraul. Res. XVth Congress, Istanbul, Turkey.*

Fleming, G. 1973. Flood frequency and the effects of catchment changes. *Wat. & wat. Engng., 77.* (926).

Fleming, G., Black, R. E. (In press). Evapo-transpiration inputs for conceptual models. *Inst. wat. Engng.*

Fleming, G., Black, R. E. (In press). Simulation of flow records for an Australian catchment using the Stanford Watershed Model IV. *Trans. Inst. Civ. Engrs. Australia.*

Fleming, G. (In press). *Deterministic simulation in hydrology*, American-Elsevier, N.Y.

Fleming, G. (In press). *Deterministic models applied to hydrology*. United Nations/FAO, Rome.

Studies of sediment bed load in parallel and converging channels
Dr J. M. Townson (M. M. Hossain) 1970-73

A comparison between numerical and physical models of parallel and converging channels uses the method of characteristics and laboratory fluxes for data representing the sediment-water interface movement.

Laboratory and numerical studies of runoff from a conceptual catchment
Dr J. M. Townson (Ong Ho Sim) 1970-72

Hydrographs resulting from a line of nozzles moving over a simple catchment have been investigated by a laboratory simulator and a numerical model based on that of Morgali (1970). Comparisons with the work of Yen and Chow (1969) have been incorporated.

Townson, J., Ong Ho Sim. (In press). A laboratory study of runoff caused by line storms moving over a conceptual catchment. *Wat. Services.*

Department of Geography (STRGG)

Infiltration rates and overland flow development on various soil types and land treatment
Dr R. C. Hills (now of East Africa Meteorological Dept., Nairobi, Kenya) 1970
Completion of study started at University of Bristol 1965.

Hills, R. C. 1971. Lateral flow under cylinder infiltrometers: a graphical correction procedure. *J. Hydrol., 13,* 153-162.

Hills, R. C. 1971. The influence of land management and soil characteristics on infiltration and the occurrence of overland flow. *J. Hydrol., 13,* 163-181.

Thermal characteristics and related hydrology of upland catchments
Dr K. Smith (Dr M. E. Lavis) 1968-70
The basic aim of this project was to investigate the fluctuations of water temperature within an upland river system and to assess the relationships with air temperature and streamflow variables. A detailed study was conducted in one small, high-level drainage basin and, at the larger scale, correlations were made at the existing flow gauging stations on the mainstream of the river Tees.

Smith, K. 1971. Some features of snow-melt recession in the upper Tees basin. *Wat. & wat. Engng., 75,* 345-346.

Smith, K., Lavis, M. E. 1972. Reservoir storage and the thermal regime of rivers, with special reference to the river Lune, Yorkshire. *The Science of the Total Environment, 1,* 81-90.

Smith, K. 1972. River water temperatures: an environmental review. *Scot. geogr. Mag., 88,* 211-220.

Chemistry of karst water
Dr S. T. Trudgill 1968-

The solution potential and dynamics of karst water are being studied.

Hillslope hydrology and water quality
Dr S. Trudgill 1974-

Development of a model of water quality based on the route water takes over and through soils/shallow bedrock and on soil-water residence time.

Curtis, L. F., Trudgill, S. 1974. The measurement of soil moisture. *Brit. geomorphol. res. Grp., Tech. Bull.*

SUNDERLAND POLYTECHNIC

Department of Civil Engineering (SNPCE)

Ripple generation on granular bed of open channel flow
Dr T. R. Davies 1973-

The formation of ripples from a plane granular bed is being studied in detail by classification of different processes observed and by use of a high-viscosity fluid to isolate the effect of laminar sublayer in a recirculating open channel 5m × 0.16m × 0.16m, using sands and glass beads.

Department of Geology (SNPGL)

Aquifer potential of Fell Sandstone of Northumberland
A. V. Hodgson and M. D. Gardiner 1969-

The hydrogeological characteristics of the sandstone formation including petrology, sorting, jointing, interbeds, porosity, laboratory permeability, chemical analysis, test pumping of existing boreholes, are being investigated in this largely undeveloped aquifer and proposals for economic development are outlined.

Hodgson, A. V. 1970. A contribution to groundwater resources in Northern England. *York. geol. Soc. Symp. Natural Resources in N. England.* (Ed: P. T. Warren). 41-46.

Hodgson, A. V., Gardiner, M. D. 1971. The aquifer potential of the Fell Sandstone of Northumberland. *Quart. J. engng. Geol., 4,* 2, 91-109.

Hodgson, A. V. 1972. A groundwater river regulation scheme. *Wat. & wat. Engng., 76,* 912, 54-58.

SURREY

Department of Civil Engineering (SURCE)

Analysis of a river regulating scheme
R. A. Scott 1970-73

The daily flow record for the River Usk was analysed as a time series to allow computer generation of extended synthetic records. These were used in simulation studies to estimate the safe yield using Usk reservoir for regulation purposes in conjunction with the pumped storage reservoir at Llandegfedd.

TEESSIDE POLYTECHNIC

Department of Civil Engineering (TWPCE)

Non-uniform aquifer flow
Dr T. Cairney (A. K. Singh) 1973-

Traditional, electrical analogue and mathematical model approaches to the Magnesian Limestone aquifer of S.E. Durham are being compared.

Cairney, T. 1972. Hydrological investigations of the Magnesian Limestone of S.E. Durham. *J. Hydrol., 16*, 323-340.

Deterioration of waters stored in flooded coal-mines
Dr.T. Cairney, Dr F. G. Butler (R. C. Frost) 1973-75

Mine waters are sampled to determine the factors influencing the mineralisation of waters in flooded collieries and to prepare planning design methods for such situations.

Cairney, T. 1973. Utilisation of disused coal-mines as water storage reservoirs. *J. Hydrol., 19*, 251-258.

INSTITUTE OF SCIENCE AND TECHNOLOGY, CARDIFF

Department of Civil Engineering and Building Technology (WISCE)

Finite element analysis of flow systems
Dr P. W. France 1974-

Investigation into methods of river flow gauging
Dr F. A. Johnson (now at SHFCE), P. H. John, P. Sutcliffe 1971-74

An experimental investigation with particular reference to less conventional methods of river flow gauging.

Water economy of Great Britain
P. Sutcliffe 1968-

An overall review involving aspects of water occurrence, conservation, use and abuse, government.

WALES: UNIVERSITY COLLEGE, ABERYSTWYTH

Department of Geography (WABGG)

Floodplain geometry
Dr J. Lewin, M. M. M. Manton (D. Hughes) 1969-76

Photogrammetric and field surveys were made of selected floodplains for the development of floodplain inundation models.

River channel processes
Dr J. Lewin (R. P. A. Brown, B. J. Brindle) 1968-75

Historical and observational data on channel forms and processes in the River Tywi have been analysed and study made of meanders confined by terrace deposits or valley walls and the role of bedforms in producing planform changes.

Lewin, J. 1972. Late-stage meander growth. *Nature, Phys. Sci., Lond., 240*, 116.

Fluvial dispersal of mining sediments
Dr J. Lewin, Dr B. E. Davies 1972-

Heavy metals in selected sedimentary environments are analysed to establish the nature and extent of mining pollution produced by river activity in association with mining processes.

Davies, B. E., Lewin, J. 1974. Chronosequences in alluvial soils with special reference to historic lead pollution in Cardiganshire, Wales. *Environmental Pollution*, *6*, 49-57.

Solutes and sediments in mid-Wales
Dr. J. Lewin (R. Cryer, D. I. Harrison, D. Grimshaw) 1970-75

Three principal lines have been followed: monitoring selected solutes in precipitation, soils and streams in upland and lowland catchments; instrumentation of catchment sediment processes, together with landform mapping and collation of data on a grid square basis, in order to develop a predictive erosion model; and analysis of sediment movement in the lower Ystwyth and Rheidd, in order to assess the possible role of such movement in the sedimentation of Aberystwyth harbour.

Lewin, J., Cryer, R., Harrison, J. I. 1974. Sources for sediments and solutes in mid-Wales. *Inst. Brit. Geogr. spec. Publ.*, *6*, 73-85.

UNIVERSITY COLLEGE OF SWANSEA

Department of Civil Engineering (WSWCE)

Numerical simulation of runoff from catchments
Dr C. Taylor (G. Al-Mashidami) 1972-

Finite element methods are being used to model direct and groundwater runoff from catchments both spatial and timewise. Variation in flow and elevation is being calculated.

Taylor, C. 1974. A finite element approach to watershed run-off. *J. Hydrol.*, *21*, 231-246.

Taylor, C., Al-Mashidami, G. 1974. Finite element solution of the shallow water equations. *Proc. F.E.M. in Flow Problems U.A.H. Press.*

Taylor, C., Al-Mashidami, G. 1974. Analysis of two dimensional surface run-off by finite elements. *TICOM Conf. on Non-Linear Mechanics Texas.*

Taylor, C., Al-Mashidami, G. 1974. A numerical technique for simulating the direct run-off component of the hydrologic cycle. *Int. Conf. F.E.M. in Engineering. Univ. of New South Wales, August 1974.*

Department of Geography (WSWGG)

Analysis of runoff and catchment characteristics in selected Breconshire (now Powys) Catchments
G. R. Hobbs, Dr D. M. Harding 1968-

Rainfall, runoff and catchment characteristics are being investigated for the catchments of the Irfon, Honddu and Upper Usk to determine the effects of the physiographic and climatic variables on runoff.

Effect of changes in the frequency of synoptic types on precipitation
Dr A. H. Perry 1971

Changes this century in the frequency of occurrence of synoptic types over the British Isles are examined and corresponding changes in the incidence of precipitation analysed.

Precipitation in Wales, in relation to different synoptic regions
Dr A. H. Perry (R. Faulkner) 1972-

Means and medians of precipitation at 17 stations in S. Wales are classified according to Lamb's synoptic weather types and the synoptic circumstances surrounding heavy daily rainfalls analysed.

Perry, A. H. (In press). A synoptic precipitation climatology of S. Wales. *Cambria* (*J. of Welsh Geogr.*)

Digital simulation of drainage basins
Dr C. Rouse (K. F. Harniman) 1969-

Drainage basins are simulated with random walk models and topological relationships and spatial patterns are analysed.

TABLE 6
ADVANCED COURSES IN HYDROLOGY AND RELATED SUBJECTS IN U.K. UNIVERSITIES

1. COURSES PREDOMINANTLY CONCERNED WITH HYDROLOGY

University and Department	*Title of Course*	*Qualification to which course leads*
BIRMINGHAM Civil Engineering	a. Water Resources Technology (with 3 options)	MSc
Geology	b. Hydrogeology	
LONDON – IMPERIAL COLLEGE Civil Engineering	Engineering Hydrology	Diploma/MSc
LONDON – UNIVERSITY COLLEGE Geology	Hydrogeology and Ground-water Resources	Diploma/MSc
NEWCASTLE Civil Engineering	a. Engineering Hydrology	Diploma/MSc
	b. Water Resources	

2. COURSES WITH SOME HYDROLOGICAL CONTENT

ASTON Biological Sciences	Biology of Water Management	MSc
CITY Civil Engineering	Hydraulics and Hydrology	MSc
BIRMINGHAM Geography	Climatology and Applied Meteorology	MSc
LONDON – CHELSEA Applied Biology	Applied Hydrobiology	MSc
LONDON – IMPERIAL COLLEGE Geology	Engineering Geology	Diploma/MSc
MANCHESTER – UMIST Civil Engineering	Water Resources	Diploma/MSc
NEWCASTLE Civil Engineering	Public Health Engineering	Diploma/MSc
READING Agriculture (in conjunction with the National College of Agricultural Engineering, Silsoe, Beds)	Agricultural Engineering	MSc
Geophysics	Meteorology	MSc
SALFORD Biology	Environmental Resources	MSc
SOUTHAMPTON Civil Engineering	Irrigation Engineering	MSc
STRATHCLYDE Civil Engineering	a. Environmental Control and Resources Utilisation	Diploma/MSc
	b. Hydraulics, Hydrology and Coastal Dynamics	

Measurement of stem flow as part of interception investigation

Institute of Hydrology

APPENDIX I

SELECTED PUBLICATIONS

PERIODICALS

Agricultural Meteorology (*Elsevier. 6 p.a.*)

Aqua (*International Water Supply Association, quarterly*)

Geographical Journal (*Royal Geographical Society, quarterly*)

Hydrological Sciences Bulletin (*International Association of Hydrological Sciences, quarterly*)

Journal of Hydrology (*North Holland, 8 p.a.*)

Journal of the Institution of Water Engineers (*8 p.a.*)

Journal of Soil Science (*Oxford University Press, quarterly*)

Nature and Resources (*Unesco, quarterly*)

Meteorological Magazine (*Meteorological Office, monthly*)

Proceedings of the Institution of Civil Engineers (*quarterly, in two parts*)

Quarterly Journal of the Royal Meteorological Society

Water Research (*International Association on Water Pollution Research, monthly*)

Water Resources Bulletin (*American Water Resources Association, 6 p.a.*)

Water Resources Research (*American Geophysical Union, 6 p.a.*)

Water Services (*formerly Water and Water Engineering; Fuel and Metallurgical Journals Ltd., monthly*)

Water Treatment and Examination (*Society for Water Treatment and Examination, quarterly*)

Weather (*Royal Meteorological Society, monthly*)

WMO Bulletin (*World Meteorological Organisation, quarterly*)

ANNUAL REPORTS AND YEAR BOOKS

Association of River Authorities Year Book and Directory (*ceased 1974*)

British Rainfall (*Meteorological Office*)

Groundwater Year Book (*Water Data Unit*)

Institute of Geological Sciences Annual Report

Institute of Hydrology Research Report

Surface Water Year Book (*Water Data Unit and Scottish Development Department*)

Water Engineers' Handbook (*from 1974 replaced by Water Services Handbook*)

Water Research Centre Annual Report (*formerly Water Research Association*)

ABSTRACTING AND INDEXING JOURNALS

Aquatic Sciences and Fisheries Abstracts (*FAO, monthly*)

GEO Abstracts B : Climatology and Hydrology (*GEO Abstracts Ltd., 6 p.a.*)

Meteorological Office Monthly Bibliography

Selected Water Resources Abstracts (*Water Resources Scientific Information Center U.S. Department of Interior, 12 p.a.*)

Soils and Fertilizers (*Commonwealth Bureau of Soils, monthly*)

WRC Information (*Water Research Centre, weekly*)

IRREGULAR SERIES

Geological Survey of Great Britain, *Memoirs*

Institute of Geological Sciences, *British Regional Geology; Well Inventory Series*

Institute of Hydrology, *Reports*

Water Research Association, *Technical Papers*

Water Resources, *Publications*

BOOKS

Barrett, E. C. and Curtis, L. F. eds. 1974. *Environmental remote sensing: applications and achievements. Papers presented at the Bristol Symposium on remote sensing, October 1972.* (Edward Arnold). 309 pp.

Carson, M. A. and Kirkby, M. J. 1972. *Hillslope form and process.* (Cambridge University Press). 475 pp.

Clark, R. T. 1973. *Mathematical models in hydrology.* (FAO, Rome, Irrigation and Drainage Paper 19). 282 pp.

Gregory, K. J. and Walling, D. E. 1973. *Drainage basin form and process: a geomorphological approach.* (Edward Arnold). 456 pp.

Institute of British Geographers. 1974. *Fluvial processes in instrumented watersheds in the British Isles.* (IBG Special Publication No. 6). 196 pp.

Institution of Civil Engineers. 1973. *Management of national and regional water resources: proceedings of the conference, London, October 1972.* 98 pp.

Institution of Water Engineers. 1972. *Symposium on advanced techniques in river basin management: the Trent model research programme, Birmingham. July 1972.* 213 pp.

Knapp, B. J. 1973. *A system for the field measurement of soil water movement.* (British Geomorphological Research Group, Technical Bulletin No. 9). 26 pp.

Monteith, J. L. 1973. *Principles of environmental physics.* (Edward Arnold). 241 pp.

Nelson, A. and Nelson, K. D. 1973. *Dictionary of water and water engineering.* (Butterworth). 271 pp.

Pereira, H. C. 1973. *Land use and water resources in temperate and tropical climates.* (Cambridge University Press). 246 pp.

Royal Society, 1971. *United Kingdom Hydrology bibliography* (annotated) 1960-1964.

Smith, K. 1972. *Water in Britain: a study in applied hydrology and resource geography.* (Macmillan). 241 pp.

Taylor, J. A. ed. 1970. *The role of water in agriculture.* (Pergamon, University College of Wales, Aberystwyth, memorandum 12). 230 pp.

Taylor, J. A. ed. 1970. *Weather economics.* (Pergamon, University College of Wales, Aberystwyth, memorandum 11). 126 pp.

Water Resources Board. 1973. *Water resources in England and Wales*. Vol. 1 : Report, WRB Publication No. 22. 67 pp. Vol. 2: Appendices, WRB Publication No. 23. 56 pp.

Young, A. 1972. *Slopes*. (Oliver and Boyd, Geomorphology Text 3). 288 pp.

GOVERNMENT PUBLICATIONS

Department of the Environment. 1971. *The future management of water in England and Wales*. A report by the Central Advisory Water Committee.

Department of the Environment. 1971. *Reorganisation of water and sewage services: Government proposals and arrangements for consultation*. Circular 92/71.

Department of the Environment. 1973. *The water services: economic and financial policies*. First report to the Secretary of State for the Environment.

Department of the Environment Management Structure Committee. 1973. *The new water industry management and structure*.

Department of the Environment and the Welsh Office. 1973. *A background to water reorganisation in England and Wales*.

A framework for Government research and development. November 1971. Cmnd. 4814 (Green Paper).

Framework for Government research and development. July 1972. Cmnd. 5046 (White Paper).

Water Act, December 1973.

Water Bill, January 1973.

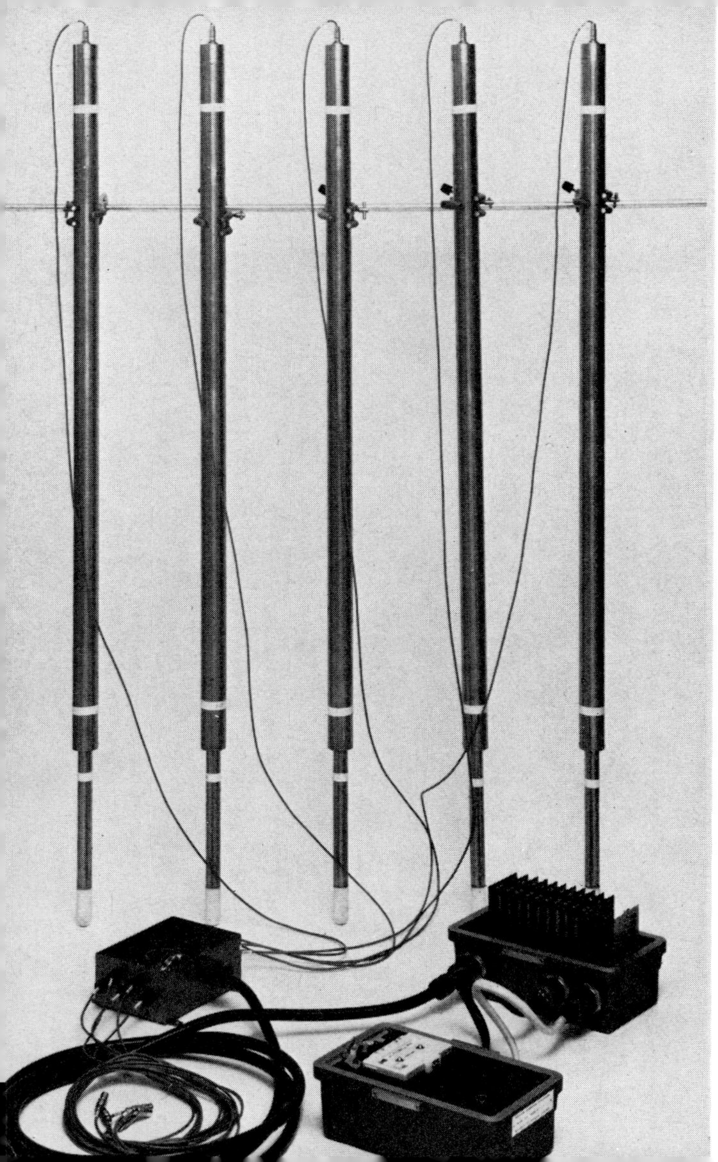

Tensiometer system for recording soil moisture tension on a Microdata logger

nstitute of Hydrology

APPENDIX II

SOCIETIES AND INSTITUTIONS

ASSOCIATION OF RIVER AUTHORITIES
Grosvenor Gardens House
Grosvenor Gardens
London SW1W 0BS
(disbanded 1974)

BRITISH GEOMORPHOLOGICAL RESEARCH GROUP
c/o Institute of British Geographers
1 Kensington Gore
London SW7 2AR

BRITISH SOCIETY OF SOIL SCIENCE
c/o University of Nottingham
School of Agriculture
Sutton Bonnington
Loughborough, Leicestershire

BRITISH WATERWORKS ASSOCIATION
34 Park Street
London W1Y 4BL

GEOLOGICAL SOCIETY
Burlington House
Piccadilly
London W1V 0JU

GLACIOLOGICAL SOCIETY
c/o Scott Polar Research Institute
Lensfield Road
Cambridge

HYDROGEOLOGICAL GROUP
c/o Geological Society
Burlington House
Piccadilly
London W1V 0JU

INSTITUTE OF BRITISH GEOGRAPHERS
1 Kensington Gore
London SW7 2AR

INSTITUTE OF WATER POLLUTION CONTROL
Ledson House
52 London Road
Maidstone
Kent

INSTITUTION OF CIVIL ENGINEERS
1-7 Great George Street
Westminster
London SW1P 3AA

INSTITUTION OF PUBLIC HEALTH ENGINEERS
32 Eccleston Square
London SW1V 1PB

INSTITUTION OF WATER ENGINEERS
6-8 Sackville Street
London W1X 1DD

ROYAL GEOGRAPHICAL SOCIETY
1 Kensington Gore
London SW7 2AR

ROYAL METEOROLOGICAL SOCIETY
Cromwell House
High Street
Bracknell
Berkshire

ROYAL SOCIETY
6 Carlton House Terrace
London SW1

INDEX TO INSTITUTIONS INVOLVED IN HYDROLOGY

FLOOD STUDIES REPORT

IN FIVE VOLUMES

published by:

Natural Environment Research Council
1975

The Flood Studies Report provides a variety of methods by which the magnitude and frequency of floods may be estimated, either at a site with flow records or at an ungauged site, and discusses the circumstances in which a particular method of estimation is suitable.

The investigations and analysis on which the recommendations are based are described, and sufficient statistical and hydrological theory is presented to provide a comprehensive guide to the hydrological, meteorological, and hydraulic aspects of flood estimation.

VOLUME I Hydrological Studies (xx + 550 A4 pages) is based on the hydrological investigations carried out at the Institute of Hydrology, and presents methods of statistical analysis of flood records, the extension of short records, the estimation of flood frequencies at ungauged sites from correlation of flood statistics with catchment characteristics at over 500 stations, the estimation of flood volumes, flood estimation by unit hydrograph techniques, and the development of further techniques of analysis.

VOLUME II Meteorological Studies (x + 81 pages), based on analysis by the Meteorological Office, describes the estimation of storm magnitudes for a given duration and return period at a given site or for a given area, with design profiles for this storm, and an estimate of maximum rainfall and snowmelt rates.

VOLUME III Flood Routing Studies (ix + 76 pages) describes the investigation of flood routing methods at the Hydraulics Research Station, and recommends methods suitable for a given situation.

VOLUME IV Hydrological Data (viii + 541 pages) presents the basic information for the gauging stations which were investigated during the study, summarises the results of unit hydrograph analysis of 1,500 storm/runoff events, and provides the basic flood records for over 500 stations in Great Britain and Ireland, supplemented by historic records.

VOLUME V Maps (24 maps) provides the maps to illustrate and apply the methods of estimation.

The FLOOD STUDIES REPORT is available at a price of £40 for the complete set of five volumes in a slip case, including packing and postage within the British Isles. (Equivalent rates are £43 for Europe and £45 for the rest of the world.) An additional set of unfolded maps, for Drawing Office use, will be supplied at a cost of £10.

For those concerned with rainfall estimates but not with river flows, Volume II (Meteorological Studies) and related maps is available separately at a cost of £15. Orders, including remittance, should be sent to the Director, Institute of Hydrology, Maclean Building, Crowmarsh Gifford, Wallingford, Oxon OX10 8BB. Cheques should be made payable to the Natural Environment Research Council.

ORDER FORM

To : The Institute of Hydrology, Maclean Building, Crowmarsh Gifford, Wallingford, Oxon, OX10 8BB

FLOOD STUDIES REPORT

I enclose the sum of £........ for .... copies of the Flood Studies Report

Date........................ Signature..................................

Full name and address for despatch :

....

....

....

....